BURNING QUESTIONS

BURNING QUESTIONS

AMERICA'S FIGHT WITH NATURE'S FIRE

DAVID CARLE

Westport, Connecticut
London

Library of Congress Cataloging-in-Publication Data

Carle, David, 1950-
Burning questions : America's fight with nature's fire/David Carle.
p. cm.
Includes bibliographical references and index.
ISBN 0–275–97371–9 (alk. paper)
1. Forest fires—United States—Prevention and control—History. 2. Prescribed burning—United States—History. 3. Forest ecology—United States. I. Title.
SD421.3.C37 2002
634.9'618'0973—dc21 2001059061

British Library Cataloguing in Publication Data is available.

Library of Congress Catalog Card Number: 2001059061
ISBN: 0–275–97371–9

First published in 2002

Praeger Publishers, 88 Post Road West, Westport, CT 06881
An imprint of Greenwood Publishing Group, Inc.
www.praeger.com

Printed in the United States of America

The paper used in this book complies with the Permanent Paper Standard issued by the National Information Standards Organization (Z39.48–1984).

10 9 8 7 6 5 4 3 2 1

Copyright Acknowledgments

The author and publisher gratefully acknowledge permission for use of the following material:

Excerpts from personal communications between the author and James K. Agee, Bruce Kilgore, Jan W. van Wagtendonk, and Paul Gleason used by permission.

Excerpts from the Harold Biswell papers used by permission of Harold Biswell, Jr.

Excerpts from Jack Ward Thomas speech at fire ecology conference in San Diego, November 2000, used by permission.

Excerpts from letters between H.H. Chapman and H.N. Wheeler, published as "Controlled Burning," *Journal of Forestry*, 39 (1941): 886–891. Used by permission.

Excerpts from letters from Wayne Hubbard to Harold Biswell, Dean Henry Vaux and Emanuel Fritz at the U.C. Berkeley School of Forestry, October 17, 1958, used by permission of Mrs. Wayne Hubbard.

Excerpts from letter from J.E. Coaldrake (Queensboro, Australia) to Harold Biswell, December 1959, used by permission of the Commonwealth Scientific and Industrial Research Organization (CSIRO).

Excerpts from letter to the editor of *National Parks Magazine*, Paul Tilden, from Harold Biswell, June 5, 1961.

Excerpts from *Proceedings, Society of American Foresters, Southern California Section, Annual Meeting, December 1, 1962, Oakland, CA* used by permission of the SAF.

Excerpts from letters from Ed Komarek and Herbert Stoddard to Harold Weaver used by permission of the Tall Timbers Research Station, Tallahassee, FL.

Excerpts from letter from Horace Marden Albright to George Hertzog, July 12, 1972, used by permission of Marian Albright Schenck.

Excerpts from letters from Newton Drury to William Penn Mott, January 9, 1973, used by permission of Save-the-Redwoods League.

Excerpts from memo from Bob Mutch to Hank DeBruin, March 10, 1972, used by permission.

Comments by Glenn Walfoort used by permission.

Every reasonable effort has been made to trace the owners of copyright materials in this book, but in some instances this has proven impossible. The author and publisher will be glad to receive information leading to more complete acknowledgments in subsequent printings of the book and in the meantime extend their apologies for any omissions.

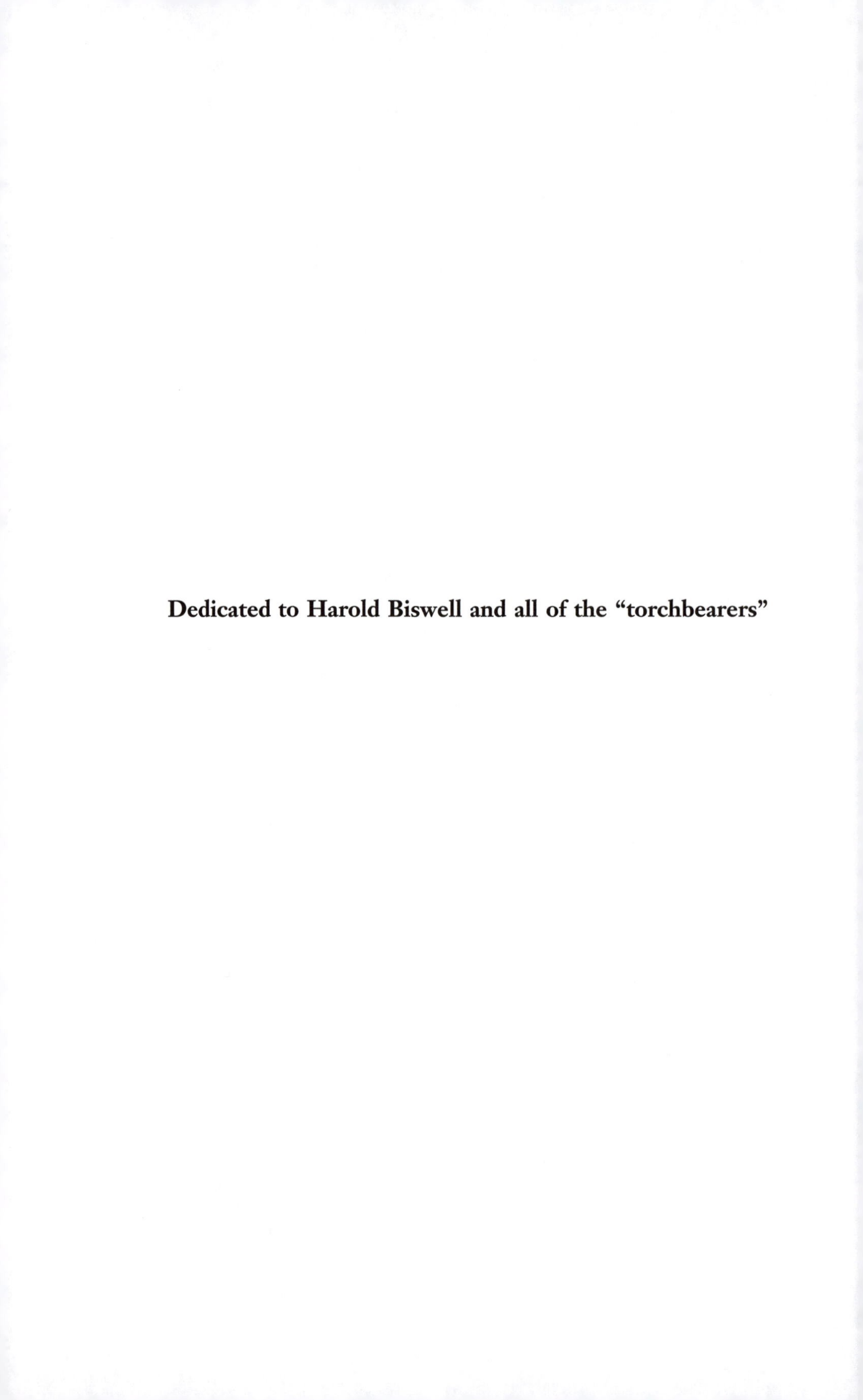

Dedicated to Harold Biswell and all of the "torchbearers"

Contents

Acknowledgments

The files and correspondence from Harold Biswell's career contain data and records important to scientific researchers following in his footsteps. They are just as valuable historically, with insights into controversies he engendered, friendships he forged with colleagues who shared similar challenges, and the personal philosophy that allowed him to endure controversy and opprobrium. After Biswell passed away in 1992, the material had been stored at his private residence. While I was researching topics for this book, Harold Biswell's son graciously gave me access to his father's files. The Biswell papers reshaped this book, allowing me to put a human face on a complex biological and sociological story. I am enormously grateful to Harold Biswell, Jr., for allowing that access. Happily, the Biswell papers found a permanent archival home at the Bancroft Library at the University of California at Berkeley during the course of this research.

I must also thank Bruce Kilgore for providing correspondence and publications from his career with the National Park Service. California State Park ecologists Jim Barry, Wayne Harrison, and Gary Fregien gave me access to files, photographs, and their personal experiences. Retired state park ranger Glenn Walfoort provided several detailed letters with recollections. Additional thanks go to Dr. James Agee (University of Washington), Jan van Wagtendonk (U.S. Geological Survey research scientist at Yosemite), Robert Mutch (retired Forest Service scientist) and Jack Ward Thomas, former Chief of the U.S. Forest Service for their comments and memories. Paul

Gleason and Roy Weaver took time to speak about a difficult episode in their professional lives, the Cerro Grande fire. Kilgore, Agee, Thomas, and Harrison reviewed the manuscript; their suggestions and corrections improved the book—any remaining errors are my own.

Archivist Cheryl Oakes, at the Forest History Society in Durham, North Carolina, was particularly helpful. As was Lisa Miller, archivist for the National Archives in San Bruno. Connie Millar (Forest Service geneticist) arranged for financial support through the Institute of Forest Genetics in Placerville, California, which made east coast research possible. My thanks to Tomás Jaehn at the Fray Angelico Chaves Library, Museum of New Mexico in Santa Fe for access to his clipping files on the Cerro Grande fire. Also to James D. Brenner of the Florida State Division of Forestry and Juanita Whiddon and Karen Gainey of the Tall Timbers Research Station in Tallahassee.

Introduction: America's Hundred Years War on Wildfire

> Keep in mind that fire is a natural part of the environment, about as important as rain and sunshine. I have heard it said . . . that we don't want to start restoring fire in our forests until we know more about its effects. . . . But isn't this a bit ridiculous when we consider that fire has always been here and everything good evolved with it? Isn't it most important to study and consider the impacts of fire exclusion? We must work more in harmony with nature, not so much against it.[1]

Dr. Harold Biswell was delivering the keynote address to the 1980 meeting of the American Association for Advancement of Science—a speech titled "Fire Ecology: Past, Present, and Future." As a professor of forestry at the University of California at Berkeley, Biswell had conducted decades of research, and his advocacy of prescribed burning on wildlands had generated fierce professional criticism. As a Forest Service researcher in the 1930s, he was introduced to California's light-burning debates; in 1940 he transferred to Georgia where controlled burns were beginning to return fire to the South's "piney woods" after four decades of exclusion policy; then from 1947 until his retirement in 1973, at UC Berkeley, his research focused more and more on the role of fire in nature. After retirement he became a consultant to park and forestry agencies that were reevaluating fire policies on landscapes where, early in the twentieth century, warfare against "evil fire" had aimed at total fire exclusion.

"What about the future of fire ecology?" Biswell asked the gathered scientists in 1980. He predicted that "Wildfire damage and number of homes burned will become much greater. Suppression costs will continue to skyrocket. Prescribed burning will become better understood and more widely used. Very slowly, but eventually, emphasis on prescribed burning will replace that on fire prevention and suppression. When it does, fire suppression will become more effective and much cheaper. There will not be as much devastation from wildfires. However, that seems many years off."[2]

A burning mix of diesel fuel and gasoline drips from handheld canisters onto the ground. Slowly a line of fire begins to creep downhill. The flames are well behaved, almost hesitant. This is a backing fire, creeping down slope and against the wind to slow and control its movement. The forest in this region has not burned for ninety years and has a heavy layer of duff and dead, fallen branches. With such a fuel load, the fire must be kept cool and tightly controlled. Where it actually moves too slowly, drip torches set small strip burns to keep the fire line even. They ensure that no fingers of fire will push ahead, then turn and burn back uphill. Even though the elements are "in prescription"—temperature, humidity, wind and fuel moisture all within acceptable limits—the character of this fire would change if it had the chance to move with the wind and upslope. That kind of fire might climb off the ground and crown into the upper gallery of tree branches, generating its own winds and an uncontrollable burn rate.

The flames might seem surprisingly docile to someone used to news stories that focus on catastrophic forest fires and the dangerous, dramatic battles to defeat them. Here rangers calmly watch the forest burn, igniting more fire here and there, squirting small streams of water from backpack containers to squelch wayward embers or fingers of flame. This kind of fire would not attract media attention unless it escaped and suddenly had to be fought. Battles are dramatic, even romantic, and always newsworthy. Prescribed burns aim for peaceful coexistence between humanity and fire. The routine of peace seldom makes the six o'clock news.

Prescriptions are sometimes designed to produce hotter fires than this. Historically, frequent, relatively cool understory fires beneath ponderosa pines kept shade-tolerant fir and incense cedar from moving in beneath the big trees and transforming the forest type. However, some trees, including the lodgepole pines that fueled much of Yellowstone National Park's 1988 fires, only release seeds from their cones after intense fires, hot enough to consume the crowns of mature trees, pass through. Though frequent low-intensity fires may burn small amounts of acreage in such forests, "stand-replacement" crown fires, on a cycle centuries long, reoccur when drought and wind conditions combine with built-up fuel. One lesson that fire ecologists learned is that no single prescription fits all habitats or conditions.

First-hand experience at a prescribed fire can be a revelation, puncturing widely held cultural misconceptions of "demon fire," as always destructive, always a danger. Since few Americans will ever have that personal experience, prescribed burns may remain mysterious and disturbingly at odds with lessons youngsters learn about the dangers of playing with matches and how important it is to be Smokey Bear's friend.

The summer fire season of the year 2000 undoubtedly added to such confusion. In May a prescribed fire at Bandelier National Monument in New Mexico escaped when unexpected high winds produced spot fires outside the burn plot. The wildfire that followed was named "Cerro Grande." It would burn 47,000 acres, enter the Los Alamos National Laboratory and, tragically, destroy 239 homes in the town of Los Alamos. Several weeks later, enormous firestorms began raging across Idaho, Montana, and Nevada. By the end of the 2000 fire season, more than 122,000 wildfires had covered 8.4 million acres, making this the nation's second-worst fire season ever. It was a summer bound to raise questions, such as "Who can we blame?"

Some politicians, with "long-term vision" firmly focused no further than their next election, cried out for quick fixes: more money for suppression equipment and manpower; more logging to, theoretically, reduce fuels in the forests; more controls over prescribed burning, a "controversial" and "dangerous" practice.

Others were more aware of a fact that had been thoroughly beaten into fire-fighting agencies for decades: a century of fire suppression—the all-out war to exterminate fire from forests and wildlands—instead produced conditions that guaranteed uncontrollable mass fire episodes across the West. The irony of Los Alamos, where high winds, low humidity, and human error brought tragedy, is that the National Park Service ignited that fire to reduce fuel loads in order to avoid such a disaster.

"Today we understand that forest fires are wholly within the control of men." That was the confident declaration of war by Gifford Pinchot,[3] who President Theodore Roosevelt had made the first chief of the United States Forest Service in 1905. Five years later, Pinchot's successor as chief forester, Henry Graves, declared that forest fire prevention "is the fundamental obligation of the Forest Service and takes precedence over all other duties and activities."[4] That policy became firmly entrenched in the coming decades and, as is common in warfare, it became heresy to express contrary opinions within that agency and within other fire control and land management agencies that followed the Forest Service's lead.

In 1910, early in the formative years for the young agency, a horrific trauma cemented the institutional approach to fire suppression. Extremely intense fires in the northern Rockies burned three million acres in Idaho and Montana (even more acreage in those states than in the year 2000). Eighty-five people were killed, seventy-eight of them fire fighters. Author Norman Maclean, writing about a 1949 fire that would also take fire fighters' lives, described "old-time rangers [that] had 1910-on-the-brain. Rangers for decades after were on the watch for fear that 1910 might start again and right in their woodpile. Some even lost their jobs because a fire got away from them."[5]

World War II brought new fears of forest fires ignited by the enemy as a wartime tactic. Fire prevention and suppression became military objectives and the patriotic concern of every citizen. The Wartime Advertising Council saw a natural role for Bambi, the deer that movie viewers had empathized with in his flight from a terrible forest fire in Walt Disney's feature movie. A year later, in 1944, the National Advertising Council created its most successful and persistent image, that of Smokey, the fire prevention bear. For over fifty years Smokey Bear's messages convinced generations of children that fire had no place in the forests.

Looking back from a twenty-first century perspective, it would be easy to assume that there were no debates about the entrenched policy. Fire exclusion dogma became so thoroughly widespread for so long that it is a surprise, now, to discover early voices of dissent. There *were* dissenters a century ago and through the decades that followed. A few far-sighted individuals spoke and wrote about alternatives to the war. In California, in the first two decades of the century, a debate was waged publicly over "light burning" versus fire exclusion. On one side were lumbermen and ranchers who had historically burned their privately held lands to control wildfire or to clear lands for grazing. The fire control bureaucracy ultimately succeeded in suppressing those light burning advocates.

In the South, "anti-war" efforts did begin changing forestry policy in the 1930s. Back in the West, at the same time that the Smokey Bear campaign was heating up, a few individuals reopened the light burning debate with solid research that supported a growing anti-war movement by the late 1960s and early 1970s. As the era's Vietnam War peace protestors could attest, opposition to wars could provoke emotional retaliation. Yet, in 1968, fire was carefully reintroduced to the Sequoia groves of Sequoia/Kings Canyon National Park. The policy change followed years of research by Dr. Harold Biswell of UC Berkeley, a patient and persistent torchbearer. In 1973 the California State Park System began to return fire to its wildland ecosystems. Prescribed burns and "let burns"—lightning fires that were allowed to burn within certain remote wilderness areas—became accepted tools in a new "total fire management" policy adopted by the Forest Service in 1978.

Yet, the transition moved slowly. From 1995 through 2000, an annual average of 1.4 million federal acres were treated with prescribed fire. Yet in January 2001 the federal land management agencies reported that 70 million acres were in critical need of fuel reduction to restore ecological integrity and avoid future high-severity fires. Another 141 million acres were at moderate levels of public safety and ecological risk due to fire exclusion and were steadily degrading toward critical conditions.[6] Without a major expansion of the effort, more decades of disastrous fires were coming.

Some of those massive fires arrived in the year 2000, the product of a century of fire suppression. Because the fire season began with an escaped prescribed fire that burned hundreds of homes in Los Alamos, many questions arose about fire policy. The questions, too often, had short-term political agendas driving them, when valid and illuminating questions could have been asked:

What ignited the Hundred Years War against nature's fire? What did professional foresters think would happen, in the long run, if fires were successfully eliminated? Who dared question those "righteous" warriors, early in the century?

Was there ever a war without a propaganda campaign? Has Smokey Bear's message damaged prospects for peace, or can Smokey now overcome a mid-life crisis?

Who were the anti-war activists of the late 1960s and 1970s? How did they persist in the face of strong opposition?

Must holocausts and "arms races" be inevitable products of war? How could the superintendent of Yellowstone National Park characterize the fires in 1988 within his park as "the greatest ecological event in the history of national parks?" What do fire ecologists now know about fire's role in nutrient cycling, in creating or maintaining wildlife habitat, in the basic health of different wildland plant communities? How are beneficial fires "prescribed" and carried out?

If, "to burn, or not to burn" is actually *not* the question, as fire suppression just increases the fuel load for inevitable future fires, what are the possibilities for an end to this Hundred Years War? Can humanity become a fire-adapted species? Will peaceful coexistence with wildland fire be our future?

This book explores a century of debates about humanity's relationship with fire—helpful servant versus dangerous master. It focuses on the stories of those "heretics" who, despite entrenched opposition, sought a better understanding of fire's natural role in the environment. H. H. Chapman of the Yale School of Forestry led the effort in the southern states; Harold Weaver, Bureau of Indian Affairs forester, pioneered out West; national conferences organized by the Tall Timbers Research Station in Florida made Ed Komarek "the father of fire ecology." Bruce Kilgore would lead the National Park System's prescribed burning program, as one of the second generation of agency personnel and academics trained or inspired by Dr. Harold Biswell in California.

Relatively little space is devoted here to the drama of thousands of wildfires fought during the last century by courageous fire fighters. Stories of such battles have, naturally, received much attention through the years. It takes nothing away from the heroic dedication of wildland fire fighters to turn the focus elsewhere. The former chief of the Forest Service, Jack Ward Thomas, put these different histories in perspective as the keynote speaker at a national fire ecology conference. Thomas told 900 fire management professionals,

> I very strongly suggest that we judge . . . actions of the past through the lenses of those who were in the game at the time. It is pointless to look at the past except to learn lessons useful in guiding present and future decisions. I suggest we ought to cease our search for scapegoats in the past and hope that those that follow us will be as kind to us in judging our actions over the next few decades. Those who hope to make use of controlled fire in ecosystem management will need the support of the old guard. They did their best to carry out the duties of fire suppression. They were incredibly successful and widely praised. They earned their reputations. They were and are proud of what they accomplished. Criticism may have had a place when change was desperately needed. The change has come . . . big time. Now care should be taken to ensure that the old firedogs know what's going on and

> their support cultivated. They must feel neither dishonored nor ignored.[7]

Differences of opinion between some "old firedogs" and prescribed fire proponents who pushed for change turned, at times, into heated drama with painfully dogmatic defiance of facts. Such war stories are recounted here not to identify scapegoats, but to uncover historic lessons that may guide the future. Most of all, this book honors a few special leaders, torchbearers whose stories have never been fully told.

The summer of 2000 showed how difficult restoring a balance to our wildlands has become after too many decades of fire suppression. Yet, if Harold Biswell were here to comment, his inveterate optimism would almost certainly still be in place. As he said in the closing remarks of his 1980 speech: "We cannot expect to undo the failures of the past . . . in a short time. That is, neglecting to understand nature and [work] in harmony with it. We must exercise great patience, and persistence, too."

NOTES

HB—Harold Biswell papers, held at the Bancroft Library, University of California, Berkeley (BANC MSS 2002/67 c).

1. Harold Biswell, "Fire Ecology: Past, Present, and Future," Keynote talk to the Ecology Section, American Association for the Advancement of Science, Davis, California, June 23, 1980, p. 1. Author's manuscript, HB.
2. Ibid., p. 9.
3. Gifford Pinchot, *The Fight for Conservation* (New York: Doubleday, Page, and Co., 1910), 45.
4. Henry S. Graves, *Protection of Forests from Fire*, U.S. Department of Agriculture, Forest Service—Bulletin 82 (Washington, D.C.: Government Printing Office, 1910), 7.
5. Norman Maclean, *Young Men & Fire* (Chicago: University of Chicago Press, 1992), 78.
6. U.S. Department of the Interior, "Review and Update of the 1995 Federal Wildland Fire Management Policy," Report to the Secretaries of the Interior, of Agriculture, of Energy, of Defense, and of Commerce, the Administrator, Environment Protection Agency, and the Director, Federal Emergency Management Agency, by an Interagency Federal Wildland Fire Policy Review Working Group (Boise, ID: National Interagency Fire Center, January 2001), ch. 1, pp. 7, 8.
7. National Fire Ecology Conference 2000, San Diego, California, November 2000.

PART I: QUESTIONING THE DOGMA OF WAR

First the idea must be seen in enthusiastic vision by someone, and enunciated for the world to hear. It must get abroad among men, and be somewhat widely considered. It must come to be deemed important. Then it must be ignored, recognized, restated, ridiculed, refuted, denied, doubted, admitted, discussed, affirmed, believed, accepted, taught to adults, taught to children, wrought into literature, put into practice, tested by its fruits, allowed to modify other ideas, embodied in institutions; and in the course of some generations it will sink in among the certainties that are assured and acted upon without question and without thought. For this process two hundred years is but a short period.

William Newton Clark

This [200 years time required for change] is not so today with radio, television, extension systems, educational and information programs.

Written out in long-hand, in the papers of
Dr. Harold Biswell

1. "Professional" Versus "Indian Forestry"

> 20 years . . . has made of complete fire protection, in all circumstances and regardless of conditions, not a theory but a religion. . . . You cannot argue with a man about his religion. He has to learn by experience.
>
> A "distinguished man of letters," 1928[1]

"In some quarters this method has been sneered at as 'the Digger Indian plan,'" the editor of the *San Francisco Call* wrote on September 23, 1902, endorsing "an experienced mountaineer's" letter. "It should be sufficient compliment to this natural method that the Indian lived in, preserved, made permanent and transmitted to us on this continent the most extensive, valuable, and useful forests in the world."

The letter from H. J. Ostrander of Merced, California, was titled, "How to Save the Forests by Use of Fire": "Scientists say that in order to preserve the forests from fire, pine needles shall be allowed to accumulate, that dead brush shall not be burned out, that fallen trees shall not be disturbed. The practical mountaineer says, 'Burn and burn often, in order that this accumulation . . . shall not become so great as to cause the destruction of the trees when a fire sweeps through the mountains.' "[2]

Ostrander's letter and the newspaper endorsement were early salvos in a public debate over light burning in California a century ago. "Light burning" referred to fires purposely ignited to control growth in forest and wildlands, keep the landscape open, favor certain plants over others, and reduce the threat of intense wildfires. Later the same practices would be called "controlled burning," a term that would then be replaced with "prescribed burning" as the practices were refined during the final decades of the twentieth century.

Controversy ignited after federal forest reserves (precursors to the national forests) were established in 1891 and efforts began to not just control wildfires, but to entirely eradicate them from forests and wildlands. It is ironic how many of the arguments used then remain familiar today. If history can teach lessons, the light burning debate over "Indian" versus "professional" forestry is history deserving contemporary attention.

By 1902 federal "forest reserves" had been established in California for eleven years. They were administered by the General Land Office of the Department of Interior. Ostrander was disturbed by those outside "scientists" laying down new fire policy and soldiers putting out forest fires. On the other hand, the reserves had been created to control logging, mining, and grazing practices across the nation that had been stripping the landscape. Loggers left behind slash—unmillable limbs and branches—that fueled increasingly large fires. Railroads provided a major ignition source for such tinder; locomotives that spewed cinders and sparks were the third cause of forest fires in 1887, after land clearing and hunters.[3]

That era's logging practices led to wildfires that killed hundreds of people in the Great Lakes states in the late nineteenth century. The most lethal wildfires in U.S. history came in 1871, originating on mainly private timberlands in Wisconsin and Michigan. The Peshtigo fire consumed that Wisconsin town in just one hour on October 8, 1871. Fifteen other towns were also engulfed, and over 1,500 people died in the regional firestorm.[4]

In California grazing practices also introduced too much unrestrained fire. In 1894 John Muir deplored the annual burning by sheep herders in the Sierra Nevada mountains: "Running fires are set everywhere with a view to clearing the ground of prostrate trunks, to facilitate the movements of the flocks and improve the pastures. The entire forest belt is thus swept and devastated from one extremity of the range to the other. . . . Indians burn off the underbrush in certain localities to facilitate deer-hunting, mountaineers and lumbermen carelessly allow their camp-fires to run, but the fires of the sheepmen, or muttoneers, form more than ninety percent of all destructive fires that range the Sierra forests."[5]

Gifford Pinchot headed the Division of Forestry within the Department of Agriculture in 1899. That October *National Geographic* printed his article, "The Relation of Forests and Forest Fires." Pinchot recognized fire as "one of the great factors which govern the distribution and character of forest growth." With an enthusiasm for a field of study that would, six decades later, be formally known as "fire ecology," Pinchot called for more understanding of "the creative action of forest fires." He acknowledged that "we have not stated everything when we say that a given forest is destroyed by fire." His intellectual curiosity was, unfortunately, constrained by an overriding belief system regarding fire. The reason to study fire's natural role in forests was just to gain "a clear and full conception of how and why fires do harm, and how best they may be prevented or extinguished. That fires do vast harm we know already." Though Pinchot identified fire as the "restraining cause" that determined the presence or absence of forests, he hastened "to add that these facts do not imply any desirability in the fires which are now devastating the West."[6]

Pinchot's journal descriptions of a tour of the West in 1891 included observations about fire and California's giant sequoias: "But who shall describe the sequoias? When the black marks of fire are sprinkled on the wonderfully deep rich ocher of the bark, the effect is brilliant beyond words. These highly decorative but equally undesirable fires bulked large in the minds of the Kaweah colonists, most of whom were Eastern tenderfeet. One of them told me they had saved the Big Trees from burning up twenty-nine times in the last five years. Which might naturally have raised the question, Who saved them during the remaining three or four thousand years of their age?"[7]

In Oregon Pinchot first encountered light burning advocates and "the belief that forest fires can be prevented only by annual burning of the leaf litter on the ground." As in California, he saw that "fire conditioned and controlled the forest in the Olympics."[8] Elsewhere he would write about a forester's need for "seeing eyes," but, though he saw the natural role of fire in those forests, for Pinchot and other foresters who shaped the early years of fire exclusion policy, such conclusions could not alter the basic wrongness of fire.

Pinchot's forestry division began a study of forest fires and their history in 1899, "reaching from Washington and Montana to Florida and Georgia, with the sound idea of finding out how much they were costing the Nation."[9] Aimed at documenting and proving that fires were destructive, the study reached a tentative figure of $20 million a year in damages.

When John B. Leiberg wrote a report on *Forest Conditions in the Northern Sierra Nevada, California*, for the U.S. Geological Survey, he too recognized fire as "the most potent factor in shaping the forests of the region. . . . In fact almost every phase of its condition, has been determined by the element of fire."[10] Yet Leiberg, in 1902, was unable to consider that a beneficial force, but rather something humanity must overcome. The land was carrying only 35 percent of the timber it was capable of supporting, he felt. Leiberg identified a region fifteen to twenty miles wide, stretching the length of the Sierra Nevada mountain range, where the forest burned historically. There fire was a thinning agent, but thinning, despite its benefits to vegetable gardens and farm crops, was not valued by government foresters.

Many private lumbermen, however, were of a different mind and resisted this new message from the government. Just before the state legislature of California convened in 1905 to consider establishing a department of forests and watersheds, forest assistant E. A. Sterling with the federal Bureau of Forestry authored "Attitude of Lumbermen Toward Forest Fires." Sterling identified insect damage and short-sighted lumbering methods as great problems, yet "certain . . . it is that fire is the greatest of forest evils." He recognized that in California forests fires typically burned as ground fires that "rarely destroy extensive stands of timber, although individual trees are severely injured and often killed." Yet his primary concern remained the wrong-headed casualness of lumbermen toward fire. He bemoaned the general attitude toward forest fires as "hopelessness, coupled in a measure with indifference. Throughout California," he wrote, "lumbermen allow [fires] to run unless they threaten their mills or are likely to spread to 'slashings' in dangerous proximity to valuable timber. This, too, is in the face of the fact that nothing is more noticeable in the Sierra forests than the burned-out bases of many of the finest sugar and yellow pines." Sterling, like Pinchot, could not see the scars on old growth trees as an indication that such trees thrived with repeated burning during their long lives.

Sterling analyzed the light burning methods of one northern California lumberman, Thomas Barlow (T. B.) Walker, who took special measures to protect the bases of individual trees before broadcast-burning, but he complained that Walker's method "leaves all young growth open to destruction and *does not get at the root of the evil*" [italics added]. Those final words are revealing—even when used as a controlled "tool," fire never was acceptable because, "at the root" it was inherently the "greatest of forest evils."[11]

In 1905 the U.S. Forest Service was born when administration of the forest reserves was transferred to the Department of Agriculture under Chief Forester Gifford Pinchot. That July field men were given copies of *The Use Book*. This manual told Forest Service employees that they had three chief duties: "To protect the reserves against fire, to assist the people in their use, and to see that they are properly used." Notably, fire protection came first on the list. The manual advised swift suppression action: "Care with small fires is the best preventive of large ones." Recognizing that fire was still widely used as an agrarian tool, forest officers were instructed to use "the utmost tact and vigilance . . . where settlers are accustomed to use fire in clearing land. Public sentiment is rightly in sympathy with home builders and the control of their operations should give the least possible cause for resentment and impatience with the reserve administration, but it should be exercised firmly none the less."[12]

Marsden Manson, San Francisco's city engineer, joined the attack on light burning practices, calling it "the Digger Indian system of forestry" in a 1906 article in the *Sierra Club Bulletin*. "Digger" was at the time a widely used derogatory term for California Indians, whose food gathering practices and "primitive" ways were scorned by that state's brash post–gold rush society. Manson, like Pinchot and Leiberg, admitted that light groundfires characterized the region's forests. "These light fires gave open forests through which one could readily see for great distances," Marsden wrote. "So impressive were these forest vistas and so majestic were the great boles that poetic and impracticable natures at once accepted the Digger Indian system of forestry as unquestionably the natural and correct one. The impression has been strengthened by . . . the absence of a definite knowledge of what forestry really is. . . . *The Digger Indian system of forestry will not give timber as a crop*" [italics added].[13]

Here was the explanation for foresters who acknowledged a historic fire regime that generated low-intensity fires, yet still saw the primeval forests as damaged by forest fires. Small trees—"reproduction"—were most susceptible to fire and had to be protected if the forests were to produce the maximum yield of the crop sought by "systematic forestry." Preserving the Sierra Nevada forests was simply a matter of fire protection and "cutting out superabundant young growth" as needed, according to Manson.

With confidence in its ability to transform the forests, war was declared on nature's fire by the young profession of American forestry. Pinchot's 1910 book, *The Fight for Conservation*, was filled with

A "light burn" surface fire on the Harney National Forest, South Dakota, in western yellow (ponderosa) pines. Photo by Henry S. Graves, 1896. U.S. Forest Service photo. Courtesy of the Forest History Society, Durham, NC.

warlike analogies, including chapters titled, "The Moral Issue," "Public Spirit," "The New Patriotism," and "The Present Battle." Regarding forest fires, he wrote, "It was assumed that they came in the natural order of things, as inevitably as the seasons or the rising and setting of the sun. *To-day we understand that forest fires are wholly within the control of men*" [italics added].[14]

Gifford Pinchot had been appointed by President Theodore Roosevelt, a Progressive and a conservationist. He was fired, in 1910, by Roosevelt's successor, President Taft, who appointed Henry Graves to replace Pinchot. In the opening sentence of a Forest Service bulletin Graves issued that year, *Protection of Forests from Fire*, he declared: "The first measure necessary for the successful practice of forestry is protection from forest fires."[15] Graves' bulletin did, however, devote a section to broadcast burning and annual burning of litter, measures he could support as long as fire was always cautiously contained. "Merely setting fire to the woods without control," Graves wrote, "is nothing less than forest destruction."

Sunset magazine became a primary media outlet, taking the California light-burning debate before the public. A corporate interest motivated *Sunset*. The magazine had been founded in 1896 by the Southern Pacific Railroad Company to promote westward travel. Its name came from the Sunset Limited train, which ran between New Orleans and Los Angeles. Railroad companies were granted federal land within California ultimately totaling 11,585,534 acres. Their acreage included large tracts of forest land, so Southern Pacific had a direct interest in forest management.

"How Fire Helps Forestry," by G. L. Hoxie, appeared in *Sunset* in August 1910. Hoxie was a civil engineer and a self-proclaimed "practical lumberman." The editorial introduction to the article reminded readers that "forest fire is a name of terror to all who love trees and who recognize the economic importance of forests. . . . it will surprise the majority of readers to learn that prevention of fire may be made so complete as to menace the forests with greater danger than they now incur. This article . . . tells how fire must be fought with fire and sounds a note of warning against the theoretical policy of the Federal Service."

"Practical foresters," Hoxie declared, "can demonstrate that from time immemorial fire has been the salvation and preservation of our California sugar and white pine forests. The practical invites the *aid* of fire as a *servant*, not as a *master*. It will surely be master in a very short time unless the Federal Government changes its ways" [emphasis in original]. Hoxie derided the "theoretical" policies of fire exclusion, worrying that if they were followed for just "a few years longer, there will be no hope of saving these areas from useless, unnecessary and enormous damage, as the accumulated fallen limbs and unusual and unnecessary hazard is many times greater in five or ten than in two or three years." The federal government threatened private land owners' livelihood, he said, because "theoretical forestry whims" were being rigidly applied on all timber lands. He added:

> The federal forest rangers—or at least many of them—know that the present theories are ruinous and calculated to insure unnecessary and great damage to the forest areas by the master fire if continued.
>
> Why not by practical forestry keep the supply of inflammable matter on the forest cover or carpet so limited by timely burning as to deprive even the lightning fires of sufficient fuel to in any manner put them

> in the position of master? . . . fires to the forests are as necessary as are crematories and cemeteries to our cities and towns; this is Nature's process for removing the dead of the forest family and for bettering conditions for the living.[16]

The summer of 1910, however, was bad timing for Hoxie's argument.

HOW DEVASTATING FIRES OF 1910 SHAPED SUPPRESSION POLICY

In August 1910 wildfires burned three million acres in Idaho and Montana. Eighty-five people were killed, seventy-eight of them fire fighters. Mining, logging, and railroad construction furnished debris piles ready to ignite; locomotives, careless loggers, and lightning set them off. Backfires, set by those fighting the blazes, added more fire to the region. All of this was intensified by a "blowup" of extreme winds.

Fire-fighting tools and techniques were primitive. Historian Hal Rothman quoted from one ranger's diary written during that fire season: "Have been fighting fire up here above Lake McDonald [presently in Glacier National Park] two days now, with nothing to work with but my hands. Skinned both of my knees climbing up here over the rocks. Both of my hands are burnt and skinned too. My God, how much longer can I stand it?" The next day's entry read: "Got the fire under control. My knees have scabbed over and feel pretty good lately, but my hands are in a hell of a shape. Damned if I'll ever fight fire with my bare hands again."[17]

John F. Preston of the Flathead National Forest remembered Bill (W. C.) McCormick who worked "on the North Fork of the Flathead during the 1910 fires. His system of firefighting in 1910 was to put out all the fires that he could by himself or with the few guards assigned to his district and then when the fires got too big, ride to . . . the nearest telephone and source of supplies, some 40 miles away, for help, gather a crew and equipment together, go back over the trail and to the fires. It is no wonder that the 1910 fires spread all over the country."[18]

The fires were widespread and killed far too many of the men who fought them. Yet the bitter defeats of that season only stiffened the resolve of those who lived through the experience. Surely, they reasoned, with more men, better equipment, more access roads and lookout towers, even fires like those in 1910 could be beaten. Pinchot, quoted by the *New York Times* on August 27, 1910, said, "If even a small fraction of the loss from the present fires had been expended in additional patrol and preventive equipment some or perhaps all of the loss could have been avoided."[19]

William Greeley, later chief forester of the Forest Service, saw his personal experience that summer as illuminating for "a young forester, thrown by chance into a critically responsible spot on a hot front. . . . I had to face the bitter lessons of defeat. . . . We had to engineer and organize the best resistance man can devise against terrific natural forces which at times are overpowering. From that time forward, 'smoke in the woods' has been my yardstick of progress in American forestry. The conviction was burned into me that fire prevention is the No. 1 job of American foresters."[20] Like the self-perpetuating cycle of hatred that can be fueled by wartime tragedies, the dogma and emotion of righteous war against fire solidified after 1910. Fire was *the* great enemy.

Congress, instead of shutting down the young Forest Service when it faced its first great failure, passed the Weeks Law of 1911 to provide $200,000 for federal-state cooperative fire control programs. Greeley moved to Washington, D.C., to implement the Weeks Law. "I was spurred on by vivid memories of blazing canyons and smoking ruins of little settlements and rows of canvas-wrapped bodies out in the Northwest. Aggressive service emissaries set forth to create a greater market for co-operative protection. . . . We were evangelists out to get converts."[21]

As fire suppression gained both emotional and financial backing, the light burning controversy kept on simmering in California. Light burning advocates, in fact, saw major wildfire seasons as bolstering *their* arguments. Secretary of the Interior Richard Ballinger (Pinchot's philosophical and political enemy in the Taft administration) told the *New York Times* that "We may find it necessary to revert to the old Indian method of burning over the forests annually at reasonable periods. One thing we intend making every effort to do is compel loggers to clean up. The litter which has been left in many places has increased the destructiveness of forest fires enormously."[22]

The two positions were becoming more polarized. "This theory of 'light burning' is especially prevalent in California and has cropped out to a very noticeable extent since the recent destructive fires in Idaho and Montana," the district forester for California wrote in June

1911. F. E. Olmsted's "Fire and the Forest—the Theory of Light Burning" ran in the *Sierra Club Bulletin*. Olmsted repeated the litany of arguments—fires killed "reproduction" and damaged forest qualities—but with even greater fervor and absolutism:

> It is said, we should follow the *savage's* example of 'burning up the woods' to a small extent in order that they may not be burnt up to a greater extent bye and bye.
>
> This is not forestry; not conservation; it is simple destruction. . . . the Government, first of all, must keep its lands producing timber crops indefinitely, and it is wholly impossible to do this without protecting, encouraging, and bringing to maturity *every* bit of natural young growth.
>
> . . . fears of future disastrous fires . . . are not well founded. Fires in the ground litter are easily controlled and put out.
>
> Fires and young trees cannot exist together. *We must, therefore, attempt to keep fire out absolutely* [italics added].

Regarding 1910, Olmsted said, "If the small fires had been properly put out at the very first there would have been no ground fires. If there had been men enough, telephones enough, roads and trails enough, they could have been extinguished and we should have had no 'crown' fires."[23]

Those arguments were repeated for a northwestern popular audience in the *Pacific Monthly*, published in Oregon. The September 1911 article by Warren E. Coman asked, "Did the Indians Protect the Forest?" Coman found the idea preposterous that Indians burned with a purpose. Generalizing inaccurately and with considerable racism, he declared that all "Indians were nomads, who dwelt in skin teepees, and banded together in tribes, whose interest were always individual and tribal and never collective, as a nation or association of tribes." They lived "always in the present," according to Coman, therefore systematic burning for long-term goals was impossible.

Coman advocated patrols to put out fires "in their incipiency," yet he was open to those same patrols using fire as a tool: "Concerted burnings in the fall by the patrol men, after the rains have begun or just before the winter snows are on the ground, will mitigate the danger."[24]

The campaign to counter light-burning advocates produced another salvo in 1912, this time in the professional *Journal of Forestry*. Richard H. Boerker called the light-burning theory

California fire prevention notice aimed at light burning advocates, 1913.

most undesirable and the most mischievous, from the standpoint of Forestry. The so-called Indian method of "light-burning" . . . was not forest protection, it was far from that, it was forest devastation pure and simple.

Strange to say, today, many of the large timber holders in California are practicing this very method to protect their mature timber from the ravages of forest fires. This is not the worst of it. They are criticizing other timber holders for not adopting these methods, and are

> influencing public sentiment in the wrong direction. It shows a woeful lack of knowledge of the fundamental principles underlying Forestry.

Fire as a tool was, again, only justified "where the fire danger is great; where there is no young growth; where the fire can be controlled; with fire-resistant species; and where the injury to the soil is justified by advantages of protection."[25]

To counter the government information campaign, T. B. Walker, the private owner whose program of light burning had been noted by Sterling back in 1904, sent his "Views on Conservation" to the San Francisco *Chronicle* in January 1913. Walker owned hundreds of thousands of acres of timber in northern California. He was convinced, the editorial introduction noted, "of the necessity of more safe and sure methods of fire protection other than the fire patrol system now in general use." Walker's extensive article dealt with objections raised by government foresters:

> I have set in operation a more complete and effective system of fire protection than any that has been heretofore used. The objections are that it kills a large part of the small sapling trees and is too expensive to apply to the vast area of the forests. As one main object is to clear away the brush and the saplings as an unquestionable necessity to provide against danger from widespread crest fires, the objection to killing undergrowth is not well taken.
>
> We are not particularly anxious to expend large sums of money unless it appears absolutely necessary and of practicable application. But we feel certain that it will prove necessary to apply a much safer method than the patrol system to insure against immeasurable damage from forest fires.
>
> The Forestry Department of the State and of the United States do not approve of this slow burning as we have been using it. Mr. Graves sees more merit in it than perhaps any other.[26]

WERE FORESTERS QUESTIONING FIRE EXCLUSION, EVEN THEN?

There were a few government foresters who attempted to find middle ground in the polarized debates over fire. Henry Graves had shown some openness to cautious use of fire as a tool. The condition of wildlands might be very different today if 1915 proposals from Forest Service staff in California had been successfully adopted. Coert duBois was then the district forester in California's District 5. Roy Headley was assistant district forester in the San Francisco office. Headley proposed "let-burn" and "light herding" policies based upon an economic approach to fire fighting. Headley, born in Illinois in 1878, had joined the Forest Service in 1907. To reduce suppression costs he was willing to let fires in brushfields burn and only conduct "loose-herding" of late fall and early spring fires, when fuels were moist and the risks were low. Low intensity fires would be allowed to spread unless they threatened high-value timber or improvements. The value of resources involved would guide suppression responses.[27] On May 1, 1915, District Forester duBois adopted this new "economic policy."

DuBois went even further when he sent Graves a draft proposal in November 1915 for a permit system to authorize private landowners to carry on controlled burns in brush. Given the debates and arguments to that point, this internal document, eventually intended as a public information circular, was amazingly honest heresy. It extolled the sought-after growth of young trees that had come in during the prior ten years of fire protection, but added that, second only to protecting the forests, it was the Forest Service's duty to assist the development of local communities. The Forest Service, duBois wrote

> would fail in this duty if it did not make a sincere effort to see clearly whether or not continued fire protection . . . might make living conditions more difficult for the mountain resident within the Forests. There is no doubt that the growth of young timber and brush has choked many old trails and made them well-nigh impassable. . . . regions where thirty years ago one could ride anywhere through the timber have of recent years so grown up with thickets of young trees and brush that cross-country travel is difficult if not impossible.
>
> It may often happen that young timber and brush may grow up so thickly in close proximity to houses, barns, or other improvements on mountain ranches or mines as to become a serious menace in case of a forest fire.

In such cases "the owners should communicate with the nearest forest officer." Informal petitions might be circulated and community requests forwarded to forest supervisors. Protective burning of strips around buildings and fields might be approved. Similarly, stockmen who could show that burning was needed to improve grazing without damage to forest resources could make a request. The Forest Service would consent to and cooperate in controlled burnings, (1) to protect homes and property, (2) to facilitate the handling of the livestock business, (3) to facilitate systematic prospecting, and (4) to clear agricultural land. They would not permit general, uncontrolled burnings of any nature. "Sincere effort has been made to see the point of view of the settler and mountain resident in this matter of brush burning, and the Forest Service . . . expects to be met in a spirit of fair play, and hopes that the policy outlined will do away with misunderstanding and friction."[28]

Chief Forester Graves applied the brakes in January, telling duBois, "It seems to me that you should not go too fast, for two reasons. You do not want the public to run away with the idea and the impression to become current—as those opposed to the principle of Government protection would doubtless be pleased to have it become current—that we have executed an about-face and thrown over our old principles. It is easy for us to draw distinctions and see distinctions where the public, or a large part of the public, will not recognize them. Again, you do not want to go faster than you can carry your own organization with you."[29] "A good point," duBois wrote here, in the margin of his copy.

Graves did not believe that the field force would know how to handle the practical problems of implementing such a policy; where to burn, how much, and when to refuse. Yet he did not kill the idea entirely, rather suggesting that duBois begin with individual projects instead of a broad public statement. "I am inclined to believe that we should go pretty slowly in taking a step which is likely to embarrass the very work which we are undertaking. Perhaps we can get up a circular on fire protection which will meet what you have in mind and also satisfy my desire to insure against losing control of the situation."

DuBois' comment, penciled along the end margin, was "A corking letter." Specific, local projects *were* instituted, including one to burn up to 15,000 acres of brush in the McCloud River area of Shasta National Forest. Forest Supervisor M. E. White submitted a proposal later that summer regarding rangelands of the Stillwater Land & Cattle Company: "Mr. R. S. Smith, the President of the Cattle Company, tells me that his stock are now coming off the range with an average of 150 pounds per head lighter than they were previous to the creation of the National Forest, when the range was burned over occasionally. . . . Burning on lands outside the Forest, directly across the

McCloud River from this area is practiced each year. Burning is recommended."[30]

Graves was reluctant to do an about-face, but knowing T. B. Walker's enthusiasm for controlled burning, convinced the wealthy lumberman to contribute $100,000 toward a chair of fire protection at the Yale School of Forestry. Graves then assigned forest examiner Stuart Bevier Show to make a fire study on Walker's property in forests of the Red River Lumber Company. Show was sent to the Feather River Experiment Station near Quincy in July 1915 and began research studies on fire spread, damage, and light burning. World War I interrupted his experiments. They were completed after 1917, in time for a forest supervisor's meeting held in Davis, California, in February 1919.

The meeting featured a debate between Show and Roy Headley over the assistant district forester's "economic" fire control policy. Show later said, "I was on to tell about the fire statistic studies, and decided I might as well be hung for a sheep as a lamb, so I put up my charts and maps and dove head on into a clear exposition of the damnable results of Headley's policies. . . . Boy, it got hot, with Headley challenging sneeringly, the Supervisors, who hated his guts, on my side, me standing pat, and duBois getting the idea, to his hatred and disillusionment."

Show convinced the supervisors and duBois. Headley "got kicked upstairs" to the Washington office and, Show said, "DuBois, for the first time, acted as though I existed."[31]

S. B. Show, known as Bevier, was born in Nebraska in 1886. His family moved to Palo Alto, California, when he was a child; his father was the first history professor at Stanford University. Show majored in botany at Stanford, then went to the Yale School of Forestry for a masters degree, completed in 1910. He worked first for the Shasta National Forest, where the following spring "a queer character, wearing a hard-boiled hat and choker collar, named Edward I. Kotok" joined the staff.[32]

Kotok was a Russian immigrant whose family settled in New Jersey when he was seven years old. He grew up in that urban environment, but after graduating from City College of New York, majoring in chemistry, he decided on a career in forestry after attending a speech by Gifford Pinchot. He earned his masters degree in forestry in 1911 and began work in the Shasta National Forest later that year. Show and Kotok worked for four years together on the Shasta and their personal bond was cemented when Show's sister, Ruth, married Kotok.

In the next decade, the brothers-in-law collaborated on several influential documents about fire in California pine forests, including, "Forest Fires in California, 1911–1920." In that document, published four years after Show's showdown with Headley, they wrote, "In principle the economic theory of protection may seem to be sound. In practice it has . . . grave weaknesses." According to their analysis, it was difficult to appraise true ultimate damage, and there was a danger that fires not attacked "with the utmost vigor may become a disaster." Most of all, there was the moral argument: "the risk that any relaxation in the speed and vigor of assault on fires may have a bad effect on discipline. Even if the theory were sound the Forest Service could not, on the one hand, urge on the public the utmost care with fire and, on the other, condemn itself for failing to follow its own preaching."[33]

Within a month of the 1919 supervisors meeting, duBois restored the earlier policy of fighting fires all-out from the start, attempting to hold all of them to the smallest size possible. The ranks closed, presenting a united front again; dogma defeated heresy and a simple, straightforward policy replaced the more complex problems associated with controlled burning. Yet the Service still faced opposition from "practical" civilians.

A retired captain with the U.S. Army engineers, Joseph A. Kitts, lived in the Sierra Nevada foothill town of Grass Valley. In 1919 he wrote an article for *The Timberman*, a West Coast trade publication for the logging industry. Kitts burned his home property following "a method of prevention of destructive forest fires learned in principle from the Sierra Nevada Indians." Assuming Kitts hoped to shape the opinions of professional foresters, he must have been tone deaf to the nuances in his description of Indian practices: "That the Indian practiced a periodic burning over of the forests is well known. He explained this to the pioneer by saying: 'Letum go too long—get too hot—killum all.' . . . we must admit that his methods of preservation and natural reproduction were highly successful when compared with our results."[34] None of that, particularly the slam at modern results, would appeal to the new scientific professionals.

Perhaps he was a poor communicator, but Kitts' rules of burning make sense today: Burn during and at the end of the wet season, at intervals of five to fifty years depending on local rates of litter accumulation. Wet humus was his key index. High winds were avoided. He backfired from barriers, toward the wind and down from tops of slopes, firing ridges before slopes and slopes before ravines, to form barriers for future burns.

STUART VERSUS STEWART

> At one of the gatherings, I tangled with Stewart Edward White on light-burning. He may have been a fine writer, but he was a lousy scientist.
>
> Stuart B. Show, 1955[35]

To seek resolution to ill feelings that threatened the cooperative fire protection programs, the Society of American Foresters arranged a series of meetings in the winter of 1919–1920. Stuart Bevier Show presented the Forest Service position at the first meeting; writer and timber owner Stewart Edward White stated the arguments for the light burners at the second.[36] White was the author of numerous novels with western settings that reflected his personal adventures in in mining and lumber camps; he also owned significant acreage in northern California's Shasta region.

The group's report to the State Board of Forestry concluded that light burning was destructive and based on false principles of forest protection and conservation. The finding was not unanimous. Committee member Ray Danaher complained that the gatherings were just to "keep the agitation out of the newspapers as much as possible."[37] The state forester agreed to conduct more experiments. Three years of study followed, monitored by a new "California Forestry Committee" whose members represented the Forest Service, state forestry, private lumber companies, and the Southern Pacific Railroad.

In the interim White prepared an article for *Sunset* magazine. Chief Forester Graves learned that the article was forthcoming and ordered a counter-campaign. "Graves Terms Light Burning 'Piute Forestry,' " *The Timberman* told its readers in the January 1920 edition. He desired to "define the issue in unequivocal terms . . . The plausible arguments advanced in support of light burning make this proposal exceptionally dangerous to our whole protective system. It weakens the confidence of the public in a genuine system of fire protection.

It weakens the united effort which all forestry interests in the Northwest have made. . . . I regard light burning as a challenge to the whole system of efficient forest protection."[38]

Graves repeated the Forest Service position that young growth was inevitably killed by fire, and special efforts to protect it were too costly. "The Forest Service has no young growth to burn up." He, instead, was pleased that "brush patches are disappearing in thickets of pine saplings." Any system that would not perpetuate forests was "purely a makeshift, which at its best is simply a part of the process of timber mining."

Graves retired early in 1920, and shortly before William B. Greeley took over in his place, Greeley's " 'Piute Forestry' or the Fallacy of Light Burning" was also published in the *The Timberman*. There he repeated, often word for word, Graves' arguments, with further amplification. Neither forester explained the term "piute," used in their article titles, but Greeley found preposterous (the same term used earlier by Coman) assertions "that the noble redskin fired the forests regularly." Just as it was "preposterous to assert that young trees can survive the process. . . . The people . . . have been taught [by the Forest Service] to believe that fire must be kept out of the woods. To a surprising degree they have recognized the truth of that slogan. Now comes an insidious doctrine that this system of fire protection . . . is unnecessary. . . . If the only solution lies in the uninterrupted destruction of young growth by light burning, we had better harvest our mature stumpage without more ado and then become a wood-importing nation."[39]

Greeley showed particular rancor with Southern Pacific for benefiting from grants of so much public land while resisting his agency's public policies.

Southern Pacific's *Sunset* magazine promoted White's article with posters in city buses proclaiming, "Your forests are in danger. The Forest Service won't save them but fire will, says Stewart Edward White in a smashing article." When the Forest Service protested to the mayor of Seattle, *Sunset* retreated, believing that the Forest Service was about to bring suit.[40]

White's article appeared in March 1920. "The general public, educated for twenty years by the Forest Service, reacts blindly and instinctively against any suggestion of fire," White wrote. "Nevertheless fire—a bad master—is an excellent servant. There are good fires and bad fires." Much of his focus was on insect damage: "To the complete satisfaction of every practical woodsman outside of the Forest Service,

"Which is preferable . . . ?" Photographs that accompanied W.B. Greeley's 1920 article in *The Timberman*, " 'Piute Forestry' or the Fallacy of Light Burning." Greeley favored the "protected" forest on the right over the "clean" forest on the left, "the ideal of the light burner." Courtesy of the Forest History Society, Durham, NC.

and *secretly to a great many inside it*, the hypothesis has been proved—Fire kills bugs" [italics added].

White addressed the Forest Service desire to let every bit of "reproduction" survive, noting, "the mature timber stand is as heavy as the soil will support. In an even stand of mature timber second growth is neither desirable nor effective. The soil is there supporting all the trees it can."

He also questioned Forest Service research conclusions based "on the false premise that every tree that showed a fire scar was doomed. If their theory were correct, there would now, after thousands of years of repeated light burning, be no trees left; and the forest stands!"

> I have had a district forester . . . inform me dogmatically that with the present perfection of his equipment and personnel he could stop any fire anywhere at any time. . . . that is what the Forest Service would like us to believe. Bunk!
>
> I have met a surprising number of men in the Forest Service who endorse [light burning] privately, though officially, of course they must "play the game."
>
> . . . one may prevent fires for five, ten, twenty-five, fifty years. But one cannot eliminate all carelessness, all cussedness, all natural causes. . . . we are painstakingly building a fire-trap that will piecemeal, but in the long run completely, defeat the very aim of fire protection itself.
>
> . . . keep firmly in mind that fires have always been in the forests, centuries and centuries before we began to meddle with them. The only question that remains is whether, after accumulating kindling by twenty years or so of "protection," we can now get rid of it safely. . . . In other words, if we try to burn it out now, will we not get a destructive fire? We have caught the bear by the tail—can we let it go?
>
> . . . in this one matter of fire in forests, the Forest Service has unconsciously veered to the attitude of defense of its theory at all costs. There is no conscious dishonesty, but there is plenty of human nature.[41]

White's powerful and eloquent arguments required the personal response of Chief Forester Henry Graves. *Sunset* provided him space in the next issue for a rebuttal titled, "The Torch in the Timber: It May Save the Lumberman's Property, But It Destroys the Forests of the Future." Graves began by bluntly declaring fire the "arch-enemy of the forest." "Mr. White," he wrote, "builds up a theory of forest protection that if put into practice would either be disastrous in its damage or unworkable on account of the cost." Again he reviewed the familiar points—that burning killed young growth and cost too much to be economical. As for its usefulness against insects, Graves insisted "Fire has no such value . . . as a remedy for bark beetle attacks, but . . . even if it did, its use would not be justified." He attacked the widespread "belief" in Indian burning: "The wisdom and foresight necessary [are] far from consistent with the relatively low stage of cultural development which the aboriginal tribes of the California mountains had reached."

To accept light burning would be "practically giving up the battle for forest perpetuation. It would mean . . . a disastrous sacrifice of all that we have gained in improved conditions through fifteen years of protection. . . . We shall not murder the patient in order to be rid of the disease."[42]

In May White explained why he was unconvinced by Graves' arguments.[43] Then *Sunset* concluded the series in the June issue with District Forester Paul Redington's "What Is the Truth?" He announced the formation of the forestry committee that included "a trained forester" whose work, he confidently expected, would mean that "light burning questions as such will rapidly recede into the background, its place being taken by the larger, more important and fundamental questions of fire damage and cost of protection."[44]

The trained forester appointed to the committee's investigations, which were to be so blatantly unobjective, was S. B. Show. In 1920 he became assistant district forester in charge of research, while Kotok took over fire control in the California district office.

The report of the California Forestry Committee, authored by Donald Bruce, forestry professor at the University of California, recorded a unanimous decision adopted January 1923. "The issue was a practical one which involved not so much the truth or fallacy of a theory as a practical and economical application of whatever truth there might be therein." Light burning was complex, "not a simple nor a single idea," because it needed to adapt to local, variable conditions. On the other hand, the fire protection system of the U.S. Forest Service "was definite, standardized and well understood. . . ." Furthermore, "under conditions where light burning seems most necessary it is too dangerous to be practicable."[45]

Concluding that the Forest Service fire protective system was more practicable and economical, the committee voted to discontinue its light burning experiments. Long-term effects to follow seemed acceptable at that time. On August 18, 1924, the California Board of Forestry followed up on the committee report by adopting a resolution condemning the practice of light burning and favoring fire exclusion.

"After 1924 light burning became an official heresy," historian Stephen Pyne wrote, summarizing the debate. "It was possible eventually to employ nearly the same practice, but not to call it by its traditional name."[46] It would only be employed once scientific studies addressed the complexities and went beyond efforts to simply demonstrate problems.

Though officially defeated, the issue and its proponents never entirely went away. They could not, since the basic flaw was with nature itself; fuel kept building up, as White and others had predicted. The bear grasped by its tail was constantly getting bigger and more powerfully dangerous.

"Having worked in the Forest and Range Experiment Station [for Kotok] from 1934 to 1940," Harold Biswell told an audience of ecologists in 1980, "and having hunted quail and dove several times with Show, I was well enough acquainted with both to know that it was not wise to mention in their presence any possible benefits from prescribed burning . . . these two people together had tremendous influence in furthering fire exclusion policies in California."[47]

Show became regional director in 1926 and served in that capacity for twenty years. In 1926 Kotok became the first director of a new forest experiment station located on the Berkeley campus of the University of California. Harold Biswell's first employment, from 1930 to 1940, was at that experiment station doing range research for the Forest Service. He had majored in zoology at Central College, Fayette, Missouri, then completed graduate studies at the University of Nebraska (a M.S. in botany and grassland ecology, then a Ph.D. in botany and forest ecology with a minor in animal ecology). Born in 1905 in the Ozark Mountains of Missouri, Biswell grew up on a 346-acre farm where he learned about soils and the benefits of thinning on crop production. Every spring his family would burn weeds and trash around the fields.[48] Biswell studied mountain meadows during his first summer in California, then put in six years on Sierra foothill woodland-grass ranges.

In 1940 Biswell moved to North Carolina, transferring to the Forest Service Southeastern Forest Experiment Station. "At that time," he said, "I looked upon fire as the arch enemy of forests."[49] But change was coming to the piney woods of the South. A separate regional battle had been underway in the South between fire suppression advocates and local burning practices. Biswell's experience there with controlled fire opened his eyes and shaped his later academic career.

NOTES

1. The "distinguished man of letters" was, perhaps, Stewart Edward White. In Cameron Jenks, *The Development of Government Forest Control in the United States* (Baltimore, MD: Johns Hopkins Press, 1928), 320.
2. H. J. Ostrander, "How to Save the Forests by Use of Fire," Letter to Editor, *San Francisco Call*, September 23, 1902, p. 6.
3. H. H. Egleston, *Report on the Relation of Railroads to Forest Supplies and Forestry*, U.S. Department of Agriculture, Forestry Division, Bulletin No. 1 (Washington, D.C.: Government Printing Office, 1887).
4. National Interagency Fire Center Historical Statistics, www.nifc.gov/stats/historicalstats.html. For more detail, see Stephen Pyne, *Fire in America, A Cultural History of Wildland and Rural Fire* (1982; reprint, Seattle: University of Washington Press, 1997), 199–211.
5. John Muir, *The Mountains of California* (1894; reprint, New York: American Museum of Natural History and Doubleday, 1961), 154.
6. Gifford Pinchot, "The relation of forests and forest fires," *National Geographic* 10 (1899) 393–403. Reprinted in *Forest History Today* (Durham, NC: Forest History Society, Spring 1999), 29.
7. Gifford Pinchot, *Breaking New Ground* (1947; reprint, Washington, D.C.: 1998), Island Press, 44.
8. Ibid., 46, 127.
9. Ibid., 144.
10. John B. Leiberg, *Forest Conditions in the Northern Sierra Nevada, California*. U.S. Department of Interior, U.S. Geological Survey, Professional Paper No. 8 (Washington D.C.: Government Printing Office, 1902), 42, 44.
11. E. A. Sterling, "Attitude of Lumbermen Toward Forest Fires," in *Yearbook of the United States Department of Agriculture, 1904* (Washington, D.C.: Government Printing Office, 1905), 133–140.
12. Pinchot, *Breaking New Ground*, 266, 277.
13. Marsden Manson, 1906. "The Effect of the Partial Suppression of Annual Forest Fires in the Sierra Nevada Mountains," *Sierra Club Bulletin* 34 (January 1906): 22–24.
14. Gifford Pinchot, *The Fight for Conservation* (New York: Doubleday, Page and Co., 1910) 45.
15. Henry S. Graves, *Protection of Forests from Fire*, U.S. Department of Agriculture, Forest Service—Bulletin 82 (Washington, D.C.: Government Printing Office, 1910), 7.

16. George L. Hoxie, "How Fire Helps Forestry: The Practical vs. the Federal Government's Theoretical Ideas," *Sunset* 34 (August 1910): 145–151.

17. Hal K. Rothman, *I'll Never Fight Fire with My Bare Hands Again* (Lawerence, KS: University Press of Kansas, 1994). Quotation is from R. I. Woesner from an unnamed ranger's diary. pp. 19, 20.

18. Ibid., 188.

19. "Pinchot Places Blame for Fires; Declares Men in Congress Who Oppose Forest Service Plans Are Responsible; Death List Is Growing," *New York Times*, August 27, 1910, p. 3.

20. William B. Greeley, *Forests and Men* (Garden City, NY: Doubleday and Co., 1951), 18, 24.

21. Ibid., 24, 25.

22. "Ballinger Talks of Fires." *New York Times*, August 26, 1970, p. 4. On the same page stories were headlined: "Fire Crisis Over, Says Forest Bureau; No Safe Death County Yet" and "Forest Fires Make Haze; Boston Reports That Smoke from West Affects Atmosphere There." For a complete discussion of the differences between Pinchot and Ballinger and the full story of the 1910 fires and their broader significance, see Stephen Pyne, *Year of the Fires* (New York: Viking Penguin, 2001).

23. Frederick E. Olmsted, "Fire and the Forest—The Theory of 'Light Burning.' " *Sierra Club Bulletin* 8 (January 1911): 42–47.

24. Warren F. Coman, "Did the Indians Protect the Forest?" *Pacific Monthly* 26, no. 3 (September 1911): 300–302, 304.

25. Richard H. Boerker, "Light Burning Versus Forest Management in Northern California," *Journal of Forestry* 10 (1912): 184–194.

26. T. B. Walker, "T. B. Walker Expresses His Views on Conservation." *San Francisco Chronicle*, January 5, 1913, p. 56.

27. Robert W. Cermak, "Fire Control in the National Forests of California, 1898–1920" (M.A. thesis, California State University, Chico, 1986), 154, 155.

28. Coert duBois, "Cooperative Brush-Burning in the California National Forests," Draft circular, 95–97–03, Box 23 (27837) "Fire, Coop. 1915–23," National Archives and Records Administration, San Bruno, California, 1915.

29. Coert duBois, "D-5, Fire Cooperation Brush Burning," Letter from Graves to Coert duBois, 95–97–03, Box 23, "Fire, Coop. 1915–23," NARA, San Bruno, California, January 24, 1916.

30. M. E. White, "Report on the McCloud River Cooperative Burning Area," 95–97–03, Box 23, "Fire, Coop. 1915–23," NARA, San Bruno, California, July 23, 1916.

31. Stuart Bevier Show, "Personal Reminiscences of a Forester, 1907–1931." Written at the request of R. E. McArdle, Chief, U.S. Forest Service, Berkeley, California, 1995, p. 94. Available in UC Berkeley Library.

32. Ibid., 10.

33. S. B. Show, and E. I. Kotok, "Forest Fires in California, 1911–1920, An Analytical Study," USDA, Department Circular 243, Washington, D.C., February 1923, p. 4. Their other collaborations included: *The Occurrence of Lightning Storms in Relation to Forest Fires in California*, USDA Weather Bureau (Washington, D.C.: Government Printing Office, 1923); *The Role of Fire in the California Pine Forest*, USDA, Department Bulletin No. 1294 (Washington, D.C.: Government Printing Office, 1924); and "Fire and the Forest (California Pine Region)," USDA, Department Circular 358, Washington, D.C., August 1925.

34. Joseph A. Kitts, "Preventing Forest Fires by Burning Litter," *The Timberman* (July 1919): 91.

35. Show, "Personal Reminiscences," p. 91.

36. Donald Bruce, "Light Burning—Report of the California Forestry Committee," *Journal of Forestry* 21 (1928): 129.

37. C. Raymond Clar, *California Government and Forestry: From Spanish Days until the Creation of the Department of Natural Resources in 1927* (Sacramento: California State Board of Forestry, 1959) 490.

38. H. S. Graves, "Graves Terms Light Burning 'Piute Forestry,' " *The Tiberman* (January 1920): 35.

39. William B. Greeley, " 'Paiute Forestry' or the Fallacy of Light Burning," *The Timberman* (March 1920): 38–39. Reprinted in *Forest History Today* (Durham, NC: Forest History Society, Spring 1999).

40. Harold K. Steen, *The U.S. Forest Service: A History* (Seattle: University of Washington Press, 1976), 136.

41. Stewart E. White, "Woodsmen, Spare Those Trees! Our Forests Are Threatened; a Plea for Protection," *Sunset, the Pacific Monthly* 44 (March 1920): 23–26, 108–117.

42. Henry Graves, 1920. "The Torch in the Timber: It May Save the Lumberman's Property, But It Destroys the Forests of the Future," *Sunset, the Pacific Monthly* 44 (April 1920): 37–40, 80–90.

43. Stewart E. White, "Getting at the Truth: Is the Forest Service Really Trying to Lay Bare the Facts of the Light-Burning Theory?" *Sunset, the Pacific Monthly* (May 1920): 62, 80–82.

44. Paul G. Redington, "What Is the Truth? Conclusion of the Light-burning Controversy," *Sunset, the Pacific Monthly* 44 (June 1920): 56–58.

45. Bruce, "Light Burning," 129–133.

46. Pyne, *Fire in America, A Cultural History*, 111.

47. Harold Biswell, "Fire Ecology: Past, Present, and Future," Keynote talk to the Ecology Section, American Association for the Advancement of Science, Davis, California, June 23, 1980, p. 5 of speech manuscript. Harold Biswell papers, held at the Bancroft Library, University of California, Berkeley (BANC MSS 2002/67 c).

48. Harold Biswell, *Prescribed Burning in California Wildlands Vegetation Management* (1989 reprint, Berkeley: University of California Press, 1999), 12.

49. Harold Biswell, "Prescribed Burning in Georgia and California Compared," *Journal of Range Management* 11, no. 6 (1958): 293.

2. Burning the Southern Woods

> During those years, the Forest Service endeavored to weed out from its publications all references to any beneficial uses of fire. In fact, control-burning had become an outlaw in the forests.
>
> Harold Biswell, 1989

The pine forests of the southern states, dominated by longleaf, slash, loblolly, and shortleaf pines, once extended west from North Carolina to eastern Texas. Longleaf was widespread when the land was settled by Europeans and became a major source of timber and turpentine. But pristine forests were destroyed by "clear-cut and get out" practices and overgrazing. Timber operators took the best logs and left the land covered with flammable logging "slash." Little seed stock remained in many areas; concerns about regeneration made fire "the red scourge of the South."[1]

Yet longleaf pine was one of the first trees recognized as being entirely dependent on fire. The key to that relationship was its pattern of root growth. For three to five years seedlings would not grow upward; foliage stayed in a "grass stage." Meanwhile, underground a strong root system developed. If there was a fire during this stage, it took off the "grassy" needles without killing the tree. Then rapid vertical growth above ground began. During that growth phase, fire became a threat, but once trees were five or six feet tall, damage from low-intensity fires was again negligible. Winter buds were protected by a thick "pubescence." Fires might kill the needles, but new ones emerged from the buds.

In Gifford Pinchot's 1899 article in *National Geographic*, he called the longleaf pine "a conspicuous and rare exception" among trees in its ability to withstand fire as a young sapling. Yet, of course, that did not stop him from promoting virtual fire exclusion. With the 1911 passage of the Weeks Act, Forest Service policies under Chief Forester Graves were extended to southern states, and a long, frustrating (for all parties) attempt began to stop the annual burning practices of rural southerners. Several professional foresters, early on, joined those who advocated controlled burning of the pine forests, but their debate was for many years kept away from the public eye. The two sides of the argument were carried by forest administrators and a cadre of evangelistic fire prevention educators versus academic researchers and a very few government research personnel who favored controlled burning.

Roland M. Harper of the Alabama Geological Survey was one of the early critics of the new fire policy. In a letter to the editor of *Literary Digest* in 1913 accompanying Harper's article, he characterized his views as "quite at variance with current traditions and teachings, no doubt . . . because most teachers and students of forestry are not familiar with the great Southern longleaf pine forests, which seem to require occasional fires for their perpetuation." Harper felt that people writing about forest fires in the northern states "where such fires are often much more spectacular and awe-inspiring than they are with us, seem to regard them as an unmitigated evil or as regrettable accidents, to be prevented by all possible means. In reality, however, fire is a part of Nature's program in this part of the world. . . . If it were possible to prevent forest fire absolutely the longleaf pine—our most useful tree—would soon become extinct." The "only just criticism" of burning as practiced in the South at that time, Harper said, was that it was done too often; "oftener than Nature intended, one might say."[2]

Harper was neither a forester nor an employee of the U.S. Forest Service, however. As an outsider he was derided as a "car-window botanist" and dismissed, according to Ashley Schiff.[3] In the book, *Fire and Water, Scientific Heresy in the Forest Service*, Schiff would detail a fascinating story of more than forty years of battle in this region to overturn fire exclusion dogma.

In California the fight was primarily between "practical" timbermen and ranchers versus "scientific foresters," but in the South a few respected scientists dared to challenge the prevailing dogma of the forestry profession with solid research that supported Pinchot and Harper's early observations.

H.H. Chapman in the Louisiana forest where his research showed the importance of fire for the reproduction of longleaf pines. Photo by W.C. Hopkins. Courtesy of Yale University Library and Carolyn Hopkins.

Herman Haupt Chapman served as a forest assistant under Pinchot from 1904 to 1906, then joined the faculty of the Yale University School of Forestry. Eventually he became head of that school and also president of the Society of American Foresters. Professor Chapman was too widely respected to be discounted or ignored. His research on fire in longleaf pines began in 1907 and convinced Chapman that periodic—but not annual—control burning was essential after young trees had several years of growth. In a 1912 *American Forestry* article, he argued against the new pressure for fire prevention in the region: "The tendency seems to be to try to pass laws modeled after those of northern states, which seek to absolutely prevent fires in the forests. . . . *But there is abundant evidence that the attempt to keep fire entirely out of southern pine lands might finally result in complete destruction of the forests*" [italics added].[4]

Chapman was so circumspect about keeping the debate within professional ranks, that it was not until 1926, nineteen years after his initial research, that he stirred up broad reaction among southern foresters. In *Bulletin 16* of the Yale University School of Forestry, Chapman showed how fire exclusion prevented longleaf seedlings from becoming established, so hardwood and other pine species were replacing the longleaf forests. He recommended control burning in longleaf pine on a three-year rotation basis, primarily to promote reproduction, but also to reduce accumulating fuel hazards and control a needle blight disease called "brown spot." Even with that bulletin, Chapman's information was not widely disseminated to foresters in the field.

Some took issue with the suppression of this information. Mississippi extension forester D. E. Lauderburn complained to Chapman in a 1931 letter that research results should be published rather than suppressed and practices standardized across the South: "Our farmers are not fools, but too many desk foresters think they are morons and not prepared to know the truth."[5]

Chapman made another effort to reach forestry professionals in 1932 with a *Journal of Forestry* article, "Some Further Relations of Fire to Longleaf Pine." He could not have been blunter about the negative consequences of fire exclusion: "On the basis of existing evidence, the writer holds the belief that if complete fire protection must be enforced on the vast areas of longleaf pine lands of the South, and is successful, *the longleaf pine will disappear as a species*" [italics in original].[6] Chapman remained careful about stirring up public controversy, though, declining to state what policy changes should be instituted. No change was forthcoming that would permit controlled burns. Even private timber operators who wanted to burn their lands were under pressure; they would lose cooperative fire prevention funding from the federal government if they violated that policy. For awhile Chapman focused on personal communications with other professionals.

Austin Cary was one of his correspondents. "No Forest Service representative in history did more to encourage and promote solid forestry in the South than Austin Cary," historian Frank Heyward, Jr., wrote. "It is possible that his thinking on [fire] was influenced by H. H. Chapman, for whom he had great admiration and who in turn respected Cary's viewpoints profoundly."[7]

Cary was born in Maine. He studied entomology and biology at Johns Hopkins and Princeton Universities from 1888 to 1891. He taught at Yale Forestry School in 1904–1905, then at Harvard until 1909. In 1910 Cary began working for the Forest Service. Special projects of his were in the Southeast as a "roving missionary," working to build good forestry practices in the turpentine woods.[8]

In 1926 Cary wrote to the extension forester of South Carolina, H. H. Tryon: "In the South . . . my mind from the beginning has been open. I do not suppose that anyone realizes more strongly than myself the vast damage uncontrolled and irresponsible fire is doing in the South, the necessity for converting the people to different practices and ideas that many hold at present. At the same time I have been open to the belief that fire might do good sometimes; the idea of employing it as a cultural measure even does not shock or repel me."[9]

Ashley Schiff quoted Cary (from a confidential source): "Putting one and two together, the question of external policy as against actual fact, comes up. That is to say, suppose you are satisfied that fire helps to promote reproduction and safeguard . . . the Florida forest, is it wise to act in accordance with that idea, or better on policy grounds to smother that belief and re-engage in the efforts to shut out all fire? I myself favor the former course." Cary felt it was bad policy to manage large government timber lands "in any other than what is believed to be the most effective and business-like fashion." And that the Forest Service should be "pleasant and friendly, easy to do business with, not rigid or stickling unnecessarily for things the local people don't believe in." Cary predicted that, "a too straightout attitude in the early years of our forest control work would later prove more or less embarrassing."[10] Yet he also kept his comments among professional ranks, unwilling to bring that embarrassment down upon them.

Cary was well liked in the profession, though a blunt-spoken nonconformist. William B. Greeley later recalled, in a fond memorial, how he warned his "bride of three weeks that her dinner guest would be a rough diamond." After dinner, in the living room, "Cary quickly shed his coat, unbuttoned his vest and stretched out on the sofa. Soon he muttered something about rubbing the hide off a heel on the last snow-shoe cruise and began fumbling with a shoe. While I struggled to restrain the mirth aroused by the arched eyebrows of the lady across the room, Cary kicked off one shoe after another, relaxed in complete comfort and began a running fire of comment on . . . doings in forestry. He was," Greeley closed, "a rugged individualist if there ever was one."[11]

Additional scientific pressure came from outside the ranks of forestry. In the South populations of bobwhite quail were in a puzzling decline, so a Cooperative Quail Study Investigation was initiated in 1924 by the U.S. Biological Survey, with private financing by owners of game plantations in Georgia and Florida. Wildlife biologist Herbert L. Stoddard headed the investigation. A one-thousand-acre plantation in Florida, known as "The Hall," was made available by Colonel L. S. Thompson. Henry Beadel, one of the financial contributors, owned Tall Timbers Plantation nearby, and he and Stoddard became close friends.

Stoddard found that fire helped quail by clearing debris that hindered feeding and movement and favoring growth of legumes whose seeds were critical to the quail diet. Stoddard's book on the research findings, *The Bobwhite Quail: Its Habits, Preservation, and Increase*, was published in 1931. He drafted the chapter concerning fire in 1928 and 1929, but had problems getting the manuscript approved for publication. Though he worked for the Biological Survey, at the time other related government agencies had review and approval rights. Stoddard's findings ran counter to the fire exclusion dogma of the Forest Service, whose fire prevention campaigns used wildlife as an emotional tool.

Stoddard later recalled: "I rewrote the fire chapter five times in the attempt to get it cleared. Finally seeing no other course to pursue, I passed the word where I knew it would spread to the effect that the fire chapter, already sadly 'watered down', would have to be cleared for publication or else I would resign and write a book on the subject that would *not be a compromise*" [italics in original].[12]

He was able to keep these statements intact in the publication: "The Bobwhite of the Southeastern United States was undoubtedly evolved in an environment that was always subject to occasional burning over. Research is greatly needed and should be carried on, for fire may well be the most important single factor in determining what animal and vegetable life will thrive in many areas."[13] He concluded that fire was justified to open forest areas for quail production, recommended winter burning, and condemned uncontrolled uses of fire.

Colonel Thompson, impressed with Stoddard's work, offered him "The Hall" plantation as a gift. Stoddard renamed it "Sherwood Plantation" and operated it as a wildlife experiment station. Years later, in 1958, Stoddard, his neighboring plantation owner, Henry Beadel, and others established the Tall Timbers Research Station, where long-term fire ecology research was conducted. With that research facility as his base, E. V. Komarek, a research assistant hired by Stod-

Herbert Stoddard in a southern pine forest. Courtesy of Tall Timbers Research Station, Tallahassee, FL.

dard in 1934, would foster an influential series of national and international fire ecology conferences in the 1960s.[14]

At about the same time that Stoddard, a wildlife biologist, was stirring up controversy over fire policy, a Bureau of Animal Industry scientist went public with his own findings. S. W. Greene had conducted six years of experiments at McNeil, Mississippi, in collaboration with the Southern Forest Experiment Station. His work showed that cattle gained more weight when grazed on burned range than on forest ranges protected from fire. Further, Greene's burning research supported Chapman's findings that longleaf pines required burned-over ground for seedlings to sprout.

Greene authored "The Forest that Fire Made," an article reluctantly published by *American Forests* magazine in October 1931 (Greene's official report would not be released until 1939, by which time he had been forced out of his job for not staying quiet).[15] Greene wrote:

> All fires in the woods are by no means forest fires that destroy useful timber, even though they are uncontrolled, and not all foresters are fanatics on the subject of fires. There are many foresters and landowners "aged in the wood," who can stand calmly by and study the effect of fires of different sorts without shouting "forest fire."
>
> Thus, we have "The Forest that Fire Made" and "The Forest that Fire Protects," for where fire is kept out for a number of years and a heavy rough of grass and pine straw accumulates, a summer fire during a dry time gives a real forest fire that actually kills saplings of good size.
>
> Forest owners may yet turn to the use of fire to fight fire and get fire insurance for the cost of a match by knowing how and when to use a match as the natives of the southern piney woods have known for generations. To these people fire was not a master but a servant.[16]

The editor of *American Forests* felt it necessary to preface Greene's article with several comments: "In this article, the author raises questions that will be warmly controverted. His conclusions . . . will come as a startling and revolutionary theory to readers schooled to the belief that fire in any form is the arch enemy of forests and forestry. Mr. Greene, it should be pointed out, is not a forester." His fifteen years with the Bureau of Animal Industry were mentioned, as were his studies of fire on forage production at the Coastal Plains Experiment Station at McNeill, Mississippi:

> His conclusions, therefore, while not official expressions of the government, have been arrived at through study and observation at first hand.
>
> In publishing this article, *American Forests* does not vouch for the accuracy of Mr. Greene's conclusions. It does, however, believe them worthy of consideration.
>
> In the meantime, the American people, so prone to be careless with fire in the woods, must distinguish between the meaning of controlled and uncontrolled forest fires. The difference is as great as that between

> the subdued fire in the hearth that warms the home and the devastating flames that reduce the home to ashes.—Editor.[17]

"Here was a new angle for foresters to think about," Frank Heyward, Jr. wrote,

> and most of them lost their heads. . . . The state foresters and Forest Service made counter charges, even entering the fields of game management and animal husbandry to do so. In retaliation Stoddard and Greene entered the field of forestry. The result was chaos. Many ridiculous statements and accusations were made. Foresters asked how quail could resist the flames of woods fires and how cattle could gain weight on soil deprived of its organic material by fires. Stoddard countered by suggesting halving . . . appropriations so as to spend a portion for controlled burning equal to that spent for fire protection. Greene steadily maintained that fires built up the fertility of the soil and . . . were necessary for longleaf reproduction.[18]

Private timber interests were, meanwhile, going ahead with controlled burning despite resistance from the Forest Service. Superior Pine Products of Fargo, Georgia, burned 30,000 to 50,000 acres annually on a private 200,000 acre forest. Their president, W. M. Oettmeier, described repercussions they experienced in the 1930s. "The Forest Service, in general, was opposed to it and at one time threatened to disallow any participation in [federal fire protection] funds if we continued.[19] The Carolina Fiber Company, near Harsville, South Carolina, conducted large-scale controlled burning after 1920. Austin Cary was a stockholder in the Sesson property, in Cogdell, Florida, and after 15,000 acres were damaged there by wildfire he pushed for thinning and controlled burning.

Finally, with the wall of silence breaking, the merits of fire in southern forests were openly addressed at the 34th Annual Meeting of the Society of American Foresters in Washington, D.C. Papers read at the meeting and the discussion that followed were presented in the January 30, 1935, edition of *Journal of Forestry* under the heading "A Tale of a Root" (reference to the fire adaptations of the longleaf pine). Chapman, then president of the society, opened the discussion portion, saying that "annual fires in the longleaf pine type have proved to be bad, but total exclusion of fire has proved to be worse."

Austin Cary commented that it was the first time he had heard certain speakers expressing views in favor of fire. "I could also, if I saw fit, remind the Southern Experiment Station of views diametrically opposed to those presented here that they long held, being very straight laced about it, as it seemed to me. For my part, in years during which I have been South, I have said as little as I could in a public way," not caring to counter the official policy "unnecessarily" and because "what one might say stood a good chance of being misunderstood or distorted when it got out into the country, with disastrous results."[20]

The need for burning and concerns about overenthusiastic reactions from the general public were hashed over. Mr. Shirley Allen of Ann Arbor, Michigan, near the end of the discussion said, "I hope you people from the South will be careful in talking to southern visitors from the Lake States. They may become enthusiastic about the merits of fire and come back home dangerous citizens. Forestry and fire won't mix in the Lake States." And Roy Headley, while stating that he had much in common with many of the papers read that afternoon, wanted to put into the record "a good natured protest over the one-sided nature of the program. If anyone is looking for evidence of censorship, for the spirit of censorship he need look no further than the make-up of the afternoon's program." That prompted a rejoinder from Ed Komarek, Stoddard's assistant on the Cooperative Quail study: "I think that this is the whole trouble. The whole program up to this time has been one-sided. This is the first time that censorship on the subject has been removed and we have been told the facts."[21]

Cary's comments about the staff of the Southern Experiment Station must have been in part aimed toward Elwood L. Demmon who directed that station. At the society meeting, Demmon presented one of the papers favoring a change in policy, with decisions about using controlled fires made, case-by-case, for individual needs and tracts of land. He still cautioned that no one should "infer from these statements that protection of forests from fire is not essential to the practice of forestry in the longleaf region."[22]

Demmon had gone through a personal transition. He looked back on that meeting in oral history comments he made in 1977: "There was a great deal of criticism of this program because here were men from the Forest Service . . . myself and others, indicating that fire could sometimes be of value. There were many foresters who said, 'Well, even if you think so, you shouldn't say so.' " He gave credit, in that interview, to Chapman's persistence: "It took Chapman and others to stimulate the Forest Service in their thinking, so that they could work out the truth of the matter. But it must be admitted that Chappy was among the first to observe many of these things and to publish the information. Chapman always seemed to enjoy a fight and he was in many of them. He was a great backer of the Forest Service in some of the fights, but if he didn't like certain Forest Service policies, he didn't hesitate to say so. He was a very energetic man of ideas who didn't hesitate to speak out."[23]

DIXIE CRUSADERS

Throughout these years of battle and change, the fire prevention campaign aimed at southerners was as intense a propaganda battle as any conceived in war. The American Forestry Association sponsored a campaign called the Southern Forestry Education Project that began in 1927 (a year after Chapman's Bulletin 16 came out). A team of "Dixie Crusaders" traveled 300,000 miles through the South in truck caravans and passed out 2 million pieces of literature. One slogan painted on their trucks read: "Stop Woods Fires—Growing Children Need Growing Trees." In three years they presented lectures and films to three million people, half of them children. Jack Thurmond, "Lecturer and Motion Picture Operator" quoted one man who came up after viewing their film in Fargo, Georgia, during the 1930 campaign: "I heard you talk and saw your show about a year ago way over in Echols County. You know, I have been burnin' my woods for more than thirty years, and after listening to you talk last year I decided that maybe it was wrong to set out fire. Now that I have seen your new picture, *Pardners*, I never expect to fire the woods again."[24]

In Dallas County, Arkansas, the campaign was led by Charles Gillette. In a 1931 article for *American Forests*, Gillette wrote, "Surely no good can come from burning the woods. Yet there are thousands of farmers who think that fire is beneficial—a part of the colossal ignorance which hangs like a millstone over the future of the South's forest lands. . . . some way must be found to remove the blindfold from the eyes of those who believe they must annually burn their forest lands." Gillette helped found the Dallas County Forestry Committee, whose efforts focused on public education. "Convincing a 'dyed-in-the-wool' woods-burner that he is doing wrong is almost a hopeless task," he found. But through meetings built around the entertainment value of motion pictures, more than one thousand people attended and "fifty timberland owners agreed to protect their timberlands from fire. . . . there has been placed in front of each of these farmers' homes a large sign stating that they are keeping fire out of their woods."[25]

The campaign was completed at the end of June 1931. "Three years—three million people," wrote W. C. McCormick, summarizing the numbers reached. "Every rural school for both the white children and Negroes, was visited in Florida and Mississippi one or more times. Nearly eighty percent of the rural schools in Georgia were visited, while the percentage in South Carolina is a little lower. Millions of pieces of literature, rulers, posters and other printed matter have been distributed. Seven motion pictures . . . were produced for the work by the project."[26]

One of the USDA films used by the Dixie Crusaders was titled *Trees of Righteousness*. The woodsburner was taught "to know the way of his transgressions" in deliberate religious tones.

The Forest Service, trying to understand the recalcitrance of the South toward fire exclusion policies, hired psychologist John P. Shea to conduct a six-month psychoanalytical study in the Blue Ridge Mountains. His 1940 report in *American Forests* was titled "Our Pappies Burned the Woods." Considering the silvicultural and ecological information that was then available, his findings seem wrongfully condescending, but they played well with foresters who still resisted policy changes.

Shea found people living at or near the "level of frustration" who insisted "what their grandfathers did was 'right.' " Fishing and hunting had been their "main pleasures," but with declining wildlife populations, "whittling and talking have become their major forms of recreation. Many fires are set to get back at outsiders, particularly officials and CCC [Civilian Conservation Corps] boys placed among them to fight woods fires." In the psychologist's opinion, woodsburning was done for simple recreation and emotional impulses. Protestations by rural residents that fires had benefits "are something more than mere ignorance. They are the defensive beliefs of a disadvantaged culture group." Forest officers needed to focus their persuasiveness on "pappies," the elder male "accepted as final authority by all blood kin and by the in-laws who live under his roof." Then they "will have won the cooperation of their numerous progeny and blood kin." They also needed to find ways to "get information accepted that otherwise flows off the backs of these people like water off the back of a duck."[27]

In 1940 young forest researcher Harold Biswell transferred from

California to the Southeastern Forest Experiment Station in Asheville, North Carolina. There he found some Forest Service researchers conducting small-scale burns, despite headquarters personnel in Washington, who still solidly resisted a policy change. Biswell's work took him, in 1941, to Brunswick Peninsula Company land on the coastal plain of Georgia, where low intensity fires were being used on 80,000 acres. "This was an eye-opening experience," he later recalled. One lone "elderly person" used roads and earlier burns to limit the spread of fire. "He had patience and much experience and he managed with full control of the flames. It was an important lesson."[28]

In 1941 Biswell began six years of prescribed burning research on the Alapaha Experimental Range a few miles east of Tifton, Georgia. Following up on Greene's controversial research, Biswell focused on the role of fire in timber production and livestock grazing.

THE SCIENTIST VERSUS THE EVANGELIST

The year 1941 also produced an exchange of letters, published in the *Journal of Forestry*, between Chapman and H. N. Wheeler.[29] Wheeler, the son of a California preacher, had been a ranger and forest supervisor in California and Colorado, the head of public relations at the Denver regional office, and from 1923 traveled as the chief lecturer of the Forest Service. He spent much of his time in the South leading a Billy Sunday evangelistic style of fire prevention campaign. Excerpts from the letters show the nature of the battle between science and "religion" in the profession of forestry (italics have been added to emphasize certain phrases). Chapman opened the exchange with a letter dated June 4:

> Dear Mr. Wheeler,
>
> Your lines and mine are so far apart that you can state, as you did, that you have yet to learn of a single forest fire in the South that did no harm, and believe it, since it is your job to arouse the people to fire consciousness and *one's effort is most effective when one allows no qualifying doubts or exceptions to mar one's convictions.* I cannot take a similar attitude as it would have led me so far astray that I would have been recreant to my trust as a scientific investigator and silviculturalist. . . . While I sometimes am tempted to regard those who adhere to fixed generalizations as bulls in a china shop, yet I hope that forestry will progress

> by the efforts of all . . . even out of efforts which sometimes appear to me as misdirected.

He repeated what Wheeler must have already known, that years of research had shown the indispensable role for fire in longleaf pine silviculture. "So please, at least, don't say you never heard of any forest fire in the South that did not do harm and merely say that you don't believe it, which is your privilege. H. H. Chapman."

Wheeler's reply was dated June 7: "I do not say that fires have not done some good. *What I do say is that they all do harm, and I firmly believe that.*" He described examples of burning in forest regions throughout the country, each time showing that the use of fire was only the lesser of two evils; that forest burning was always, still, an evil. As for fuel reduction to prevent holocaust fires, he answered such arguments "by saying that it matters little whether we destroy the little trees every year or only have a hotter fire once in several years." He considered the Southern Forest Experiment Station's findings that fire benefited pure longleaf pine stands unimportant, as "there are so few stands of pure longleaf pine. . . . Publicity of supposed findings on the McNeil experiment [Greene's research] has been very detrimental to fire prevention in the South." He further stated:

> I fully realize that my lecture methods to put over a national program of forestry, and especially to stop woods burning, is not approved by all foresters, particularly those engaged in research . . . but I see no way to put over forestry in general except *to carry on, not entirely in a purely intellectual manner, but somewhat in the nature of a crusade.*
>
> When we have finally scotched this enemy fire, perhaps we can begin to talk and write about it and practice controlled burning in some localities. The present need is that all agencies, scientists and laymen alike put forth every bit of energy in fighting the forest fire menace and *do nothing to give encouragement to woods burners in any section of the United States.* We must fight the fire bug as we would a foreign enemy . . . a national defense measure of the first magnitude. H. N. Wheeler.

On June 11 Chapman answered: "I do not wish to try to undermine your belief in the infallibility of the doctrine of complete fire exclusion, which *represents so closely the official attitude at Washington that the forest supervisors and rangers have no choice but to take this attitude* and are judged by their success in preventing all fires." But policy was being translated into suppression of facts, Chapman complained. He pointed to the spring meeting of the Society of American Foresters in Lufkin, Texas, where an excursion was routed "so as to avoid an important and instructive demonstration of experimental burning in longleaf pine by the Forest Service" because the organizers "did not wish to be responsible for any encouragement of the use of fire in Texas." Chapman questioned professional misrepresentations and

> the belief that the whole truth if told will do more harm than good. . . . The cure in my mind for misleading publicity derived from experimental use of fire . . . is not more misleading publicity based on the premise that all fires are harmful or on the suppression and concealment of the facts. It lies rather in open and honest publicity to counteract misstatements of both kinds.
>
> If publicity of proper use of controlled fire is to be permanently banned by the profession or by that element within it which has to meet the responsibility of controlling the public, then I see no hope of ever actually controlling fire in the South

because informed land owners would go their own way and because fire exclusion would inevitably produce uncontrollable fire risks, as fuels built.

Wheeler came back on June 30:

> Dear Professor Chapman,
>
> You speak of facts on controlled burning. Can you be sure that what seem to be facts after a few years of experimentation may not prove to be fallacies in the long run?
>
> Now if it were proven over a long period of time, and I would say it would take 50 to 100 years to be sure of it, that woods burning is helpful under some conditions, I would not hesitate to say so before at least some of the audiences that I reach.
>
> It was impossible for me to attend the meeting of the Society of American Foresters in Washington in 1935 when controlled burning was freely discussed. My information is that the controlled burners had unrestricted opportunity to tell all they wished about it and no one took the floor in refutation.
>
> Because some of us are unwilling to accept the so-called facts about controlled woods burning until experiments have been carried on for a much longer period must we be called ignorant?
>
> You and I will not live long enough to know whether controlled burning does more good than harm. You apparently think your methods will reduce uncontrolled fires, and I am confident it will increase them, if the information is spread far and wide over the country. H. N. Wheeler

Chapman wrapped up the series of letters on July 2: "Dear Mr. Wheeler: . . . we have the internal conflict between those who would use research as a means of postponing action, and those who are willing to accept all available evidence, research included, in determining policies rather than permit resources to be damaged or wasted while awaiting absolute confirmation of unpalatable facts. H. H. Chapman."

It is probable that Wheeler's close-minded obstinacy, so openly revealed in these letters, helped convince some forestry professionals that old attitudes were an embarrassment and needed to change. Wheeler's adamancy might be easier to understand given the broader national context. The nationwide war against wildfires was given a new directive in 1935 by Chief Forester Gus Silcox (another veteran of the 1910 holocausts in the Rockies). Silcox declared that forest fires were "wholly preventable" when he issued the "10 AM Policy" for fire fighters. The goal would now be control of wildfires by 10 o'clock the morning after their discovery; if that goal could not be met, then 10 A.M. the next day became the new objective. All fires were to be hit hard and fast to be kept as small as possible.

The belief that total fire exclusion only required sufficient manpower and equipment was given its test, throughout the nation from 1934 to 1942. President Franklin Roosevelt's New Deal programs to battle the Great Depression included the Civilian Conservation Corps (CCC) which put three million men to work in the national forests and parks. Just within the forests, during the seven years of the program, CCC crews built 3,470 fire towers, installed 65,100 miles of telephone lines, thousands of miles of firebreaks, roads and trails (including 97,000 miles of truck trails and roads), and devoted 4.1 million man-hours to fighting fires.[30]

Yet uncontrollable holocaust fires, the product of built-up fuels, were the factor that finally convinced the Forest Service to change policy within the southern states. A fire on the Ocala National Forest in Florida burned 30,000 acres in 1935; 13,000 acres burned in 1941's Impassable Bay fire on the Osceola National Forest; and 1943 added the 9,000-acre Mt. Carrie burn.

The turn-around year was 1943. During the middle of World War II, a "treaty" was signed declaring a regional "truce" (though not the end of the national war against fire). Historian Schiff called the reversal of policy, "The Switch in Time that Saved the Pine."[31]

Before a meeting to consider the South's policy change, a letter from a county agricultural agent was circulated among state foresters, hoping to shape their opinions: "I have come to the conclusion that our efforts . . . have been practically worthless. I see no improvement in preventing forest fires. . . . There is something radically wrong with a procedure that does not get any better results in a quarter of a century."[32] On August 3, 1943, Chief Forester Lyle Watts authorized controlled burning in National Forests with longleaf and slash pine. In December that authorization was actually implemented when an information statement was finalized at the headquarters of the Ocala National Forest in Lake City, Florida. It became known within fire management agencies as the "Treaty of Lake City."

Peace treaties do not in themselves insure that committed warriors will lay down their arms. C. F. Evans, one of Chapman's early supporters among foresters, wrote an article for *American Forestry* in 1944 that seemed to be aimed at the minds of those fire fighters who were reluctant to embrace the change. His title sounded like the old call to arms: "Can the South Conquer the Fire Scourge?" Opening with the reassuring (for the old guard) and familiar message that the South had too much incendiarism and a need for effective fire control, Evans gradually worked toward another message—an apology to those southern "pappies" whose folk wisdom had been under so much attack. "We have been too reluctant to listen to the native who used fire to serve his purposes and to learn from his experience. He has always used fire as a means to an end and, in his viewpoint, fire was not the evil we proclaimed it to be. He used it unwisely in most cases, to be sure, but there was more validity to his practices than appeared on the surface."[33]

So, by April, 1946, 580,000 southern acres were burned under the new policy. H. H. Chapman, called by Ed Komarek "the father of controlled burning for silvicultural purposes in this country,"[34] looked back at four decades of controversy and saw "a fundamental lack of trust in the innate intelligence of farm and forest owners and workers, in the belief that any tolerance of fire in the forest will cause . . . the situation to get completely out of hand."

Chapman added that "In the long run such a policy of deliberately repudiating or concealing known facts is bound to fail and to bring permanent discredit upon those who advocate it and upon the profession they represent. It is unsound scientifically, professionally, and from the standpoint of psychology and public cooperation." He termed forest fire control and prescribed burning, "Siamese twins": "The one cannot continue to thrive without the other, and quarrels between Siamese twins are both uncomfortable and unprofitable."[35]

NOTES

1. E. V. Komarek, "Comments on the History of Controlled Burning in the Southern United States," in *Proceedings, 17th Annual Arizona Watershed Symposium*, Arizona Water Commission Report No. 5, Phoenix, September 19, 1973.

2. Roland M. Harper, "A Defense of Forest Fires," *Literary Digest*, August 9, 1913.

3. Ashley Schiff, *Fire and Water, Scientific Heresy in the Forest Service* (Cambridge, MA: Harvard University Press, 1962), 25.

4. H. H. Chapman, "Forest Fires and Forestry in the Southern States," *American Forestry* 18 (August 1912): 512.

5. Schiff, *Fire and Water*, 53.

6. H. H. Chapman, "Some Further Relations of Fire to Longleaf Pine," *Journal of Forestry* 30 (1932): 603.

7. Frank Heyward, Jr., "Austin Cary, Yankee Peddler in Forestry," Part 2 (first part in May 1955 issue), *American Forests* 62 (June 1955): 28, 29.

8. William B. Greeley, "Austin Cary as I Knew Him," *American Forests* (May 1955): 30.

9. Heyward, "Austin Cary," 29.

10. Schiff, *Fire and Water*, 40.

11. Greeley, "Austin Cary," 30.

12. H. L. Stoddard, "Use of Fire in Pine Forests and Game Lands of the Deep Southeast," in *Proceedings, First Tall Timbers Fire Ecology Conference* (Tallahassee, FL: Tall Timbers Research Institute, 1962), 31–42.

13. H. L. Stoddard, *The Bobwhite Quail, Its Habits, Preservation and Increase* (New York: Scribner, 1931) 401, 402.

14. For history of the Tall Timbers Research Station, see E. V. Komarek's articles: "The Use of Fire: An Historical Background," in *Proceedings First Annual Tall Timbers Fire Ecology Conference* (Tallahassee, FL: Tall Timbers Research Institute, 1962), 7; "Comments on the History of Controlled Burning in the Southern United States," in *Proceedings, 17th Annual Arizona Watershed Symposium, September 19, 1973*, Arizona Water Commission Report No. 5. Phoenix, September 19, 1973; and "A Quest for Ecological Understanding: The Secretary's Review, March 15, 1958–June 30, 1975" (Tallahassee, FL: Tall Timbers Research Station, 1977), 12–16.

15. W. G. Wahlenberg, S. W. Greene, and H. R. Reed, "Effects of Fire and Cattle Grazing on Longleaf Pine Lands, As Studied at McNeill, Miss.," Technical Bulletin No. 683, U.S. Department of Agriculture, Washington, D.C., June 1939.

16. S. W. Greene, "The Forest that Fire Made," *American Forests* (October 1931): 583, 618.

17. Ibid., 583.

18. Frank Heyward, Jr., "History of Forest Fires in the South," *Forest Farmer* 9, no. 8 (1950): 10.

19. W. M. Oettmeier, 1956. "The Place of Prescribed Burning," *Forest Farmer* (May 1956): 6.

20. Society of American Foresters, "Report of the 34th Annual Meeting of the Society of American Foresters," *Journal of Forestry* 38 (1935): 357. Austin Cary retired in 1935; he died in 1936.

21. Ibid., 360.

22. E. L. Demmon, "The Silvicultural Aspects of the Forest-Fire Problem in the Longleaf Pine Region," in Society of American Foresters (1935), "Report of the 34th Annual Meeting of the Society of American Fosters." *Journal of Forestry* 38: 330.

23. Elwood R. Maunder (interviewer). "Voices from the South: Recollections of Four Foresters." Oral history interviews with Inman F. Eldredge, Walter J. Damtoft, Elwood L. Demmon, and Clinton H. Coulter. Forest History Society. Santa Cruz, California, 1977, pp. 135, 136.

24. Jack Thurmond, "Through 1930 with the Dixie Crusaders," *American Forests* (March 1930): 151.

25. Charles A. Gillette, "Campaigning Against Forest Fires," *American Forests* (April 1931): 209, 256.

26. W. C. McCormick, "The Three Million," *American Forests* (August 1931): 479–480.

27. John P. Shea, "Our Pappies Burned the Woods," *American Forests* (April 1940): 159–174.

28. Harold Biswell, *Prescribed Burning in California Wildlands Vegetation Management* (1989, reprint, Berkeley: University of California Press, 1999), 13.

29. H. H. Chapman and H. N. Wheeler, "Controlled Burning," *Journal of Forestry* 39 (1941): 886–891.

30. T. H. Watkins, *The Great Depression* (Boston: Little, Brown & Co., 1993), 131.

31. Schiff, *Fire and Water*, 95.

32. Ibid., 97, 98.

33. C. F. Evans, "Can the South Conquer the Fire Scourge?" *American Forestry* 50 (May 1944): 229.

34. Komarek, "Comments on the History," p. 14.

35. H. H. Chapman, "Prescribed Burning Versus Public Forest Fire Services," *Journal of Forestry* 45 (1947): 808.

3. Harolds of Change

HAROLD: "leader of the army" from Old English haer "army" and weald "leader."

On May 1, 1947, Harold Biswell took a position at the University of California at Berkeley, teaching in the School of Forestry and conducting research in plant ecology through the University Experiment Station. Working as associate professor of forestry and associate plant ecologist meant an immediate sacrifice of $1,700 in annual salary (with the U.S. Forest Service he had been paid $7,100 per year).[1] He explained the career move in his resignation letter to the director of the Southeastern Forest Experiment Station, saying that he was leaving "with a great deal of hesitation." He took pride in the Forest Service research organization, liked the job, and felt that the agency had been good to him. "For several years, however, I have wondered how teaching might be and thought this would be a good opportunity to give it a trial. In the new job I will still have a hand in Forest Service work, indirectly, of course—that of training potential candidates. At least, I hope some of our students will be trained well enough to be able to get into Forest Service work!"[2]

Rather then traveling straight to California, Biswell stopped in Washington, D.C., "to pay respects to certain foresters there whom I knew. E. I. Kotok, chief of research at that time, said, 'Now, when you go to California, don't let them get you involved in research on control burning. Stay out of it and work on grazing problems in the high mountain meadows.' I thought this good advice because I knew something about the controversy surrounding fire and figured that I would be continually hampered if I got involved."[3]

This was the same Kotok, of course, who, with his brother-in-law S. B. Show, tamped out the light-burning debates in California and helped shape fire control policy afterward. Kotok had been Biswell's supervisor during his first years of research work with the Forest Service in California.

In 1945 the California legislature authorized the Division of Forestry to issue controlled burning permits for brush-range improvement, which led to funding for university research into the use of fire for wildlife range improvement. Despite Kotok's advice and the controversy controlled burning research would generate, Dean Walter Mulford of the School of Forestry told Biswell to "develop sound research, let the chips fall where they may, and not argue with people but rather listen to them and present facts."[4] He followed that advice throughout a tumultuous and influential career at the university that would last twenty-six years.

Biswell first taught range management courses. He soon began studying the use of fire in Sierra foothill woodlands to improve ranges for livestock grazing, along with a 65,000-acre project burning chaparral shrublands in Lake County, north of San Francisco for game habitat improvement. The fact that he was scientifically prying open a door that had been slammed shut in California for several decades did not go unnoticed. Professor Robert F. Griggs wrote him in October 1949 from the University of Pittsburgh congratulating him on his first published reports of the effects of brush burning that increased surface flow in nearby streams: "I often wonder why with all the controversy indulged in past years nobody undertook careful detailed studies such as you are carrying forward. You are much to be congratulated on the way you are carrying forward."[5]

In the early 1950s he developed a method of up-slope burning in chaparral without building firelines. South-facing slopes were ignited at the bottom of steep slopes in the spring; the fires would go out when they reached the wetter north-facing vegetation at the ridgetops. Biswell would later recall that "the California Department of Fish and Game liked it so well that they carried on for another ten years. The burning was then stopped because of money shortages and largely forgotten. The method was rediscovered by the Forest Service in the early seventies and is now gaining in popularity."[6]

He started burning in ponderosa pine in 1951 on the Teaford For-

Dr. Harold Biswell conducting a demonstration prescribed burn. Photo by Michael Yost.

est in the Sierra Nevada foothills and in the fall of that year at Hoberg's Resort in Lake County. Ponderosa pine was the most widespread forest type in western North America, with about 36 million acres from British Columbia down to Mexico. California had about four million acres of ponderosa pine forest. Biswell's interest in burning in western pine forests had been piqued while he was still in Georgia, when he read a 1943 *Journal of Forestry* article written by Harold Weaver, titled "Fire as an Ecological Factor in the Ponderosa Pine Region of the Pacific Slope."

HAROLD WEAVER

In 1943 forester Harold Weaver had been working for fifteen years on the Indian reservations of the Pacific coast. Born in Sumter, Oregon, he graduated with honors as a forester from Oregon State College, Corvallis, in 1928 and that same year began a career with the U.S. Department of Interior's Bureau of Indian Affairs (BIA). In 1933 he transferred to the regional office in Spokane, in charge of Civilian Conservation Corps work. From 1940 to 1948 he worked on the Colville Reservation in Washington, where in 1942 he began conducting prescribed burns. Weaver burned logging debris or "slash" and also extended broadcast burning to forest areas dense with ponderosa pine saplings. Broadcast burning was done downslope against prevailing winds in the autumn. The fire thinned thickets from 2,430 stems per acre to 690, and "crop trees" responded to the reduced competition with greater diameter growth in the following years than trees in nearby unthinned plots. Weaver's first article on the research was published during World War II. He explained that the dense stagnating stands that had developed on vast areas due to fire exclusion "aggravated . . . beetle losses tremendously as a result of their competition with the larger trees for the limited soil moisture" and that fire hazard had increased tremendously. "Fires, when they do occur, are exceedingly hot and destructive and are turning extensive areas of forest into brush fields."[7]

When he submitted his first article to the *Journal of Forestry* in May 1942, Weaver wrote to the editor that he was "very naturally . . . extremely hesitant to advance any suggestion or argument in favor of any type of burning, appreciating that [I] might easily be charged with advocating burning . . . when it was necessary from an unenlightened public viewpoint to discourage anything that savored of burning, controlled or otherwise. However," Weaver added, "it would appear that forestry has reached the stage where it is accepted in all intelligent quarters as a sound, progressive and unbiased profession." The paper, he wrote, was submitted to encourage "intelligent and thoughtful debate."[8]

Despite Weaver's optimism about the progressive state of the profession, his agency, the BIA, would not officially endorse the paper and required him to include a disclaimer: "This article represents the author's views only and is not to be regarded in any way as an expression of the attitude of the Indian Service on the subject discussed."

Because of the controversial subject, additional comments were appended to the end of the 1943 article by Arthur A. Brown, associate editor of the *Journal of Forestry*. "Mr. Weaver offers some challenges to fire control policies of public agencies," Brown wrote, "that deserve careful consideration." He noted that, "In California much research has been conducted on how to get ponderosa-pine reproduction rather than how to get rid of it," questioning the applicability of Weaver's concerns outside of eastern Oregon, but did close by strongly agreeing with Weaver's call for more research.

Before making the decision to publish such an article, the editor first had it reviewed by Duncan Dunning, a researcher at the California Forest and Range Experiment Station. During that period, forest entomologist Paul Keen, Weaver's mentor and good friend, wrote Dunning about the paper: "I may be charged with the responsibility of having started Weaver thinking along these lines, for I am convinced that we need a great deal more careful research on this important problem. Instead of this being a closed chapter, as some foresters would like to believe, I find an increasing skepticism on the part of many foresters as to the ecological and biological soundness of present fire control policy in the ponderosa pine region. . . . This issue can never be closed until the truth is brought to light. . . . Weaver's article is a plea for this much-needed research work and open-mindedness."

Keen described the "stormy course" of Weaver's paper, finished two years earlier then run through a gauntlet of Department of Interior officers and critics at the Forest Experiment Station in Portland, Oregon. Keen was obviously concerned about a negative reaction based solely on controversy and urged Dunning's favorable support of the paper, saying "I do know that you are a defender of the truth in research and have no sympathy with any policy of suppressing truth for administrative convenience."[9]

After the article appeared, Keen sent his congratulations: "Dear Hal: Well, I see you hit the spot-light in the January *Journal of Forestry*; now you had better begin looking for a good safe bomb shelter. As you say, if they are all as easy to answer as Brown, there will be nothing to it. His statement [in the comments following the article] that fire fighters can't imagine it being done shows that fire fighters haven't much imagination, for your controlled burning experiments show that you've done it."[10]

Keen also passed along to Weaver a memorandum he received from F. C. Craighead, another Forest Service entomologist (at the Beltsville Research Center in Maryland). "I am pleased to see Weaver express these ideas," Craighead said. "It will start something. It's too bad that someone from the Forest Service could not have done it earlier. . . . Everybody just about admits that prolonged protection is good for just so long then when conditions become right the whole thing explodes—killing everything. The most ardent protectionists have lost faith following these [large wildfires]. From my contacts, I believe there are many men in the Forest Service who feel the same way. The period of transition is going to be difficult."[11]

One forest supervisor, Percy E. Melis, from the Clearwater National Forest in Idaho, used congratulatory language in his letter that actually reveals how great the professional risk was for Weaver: "It takes a lot of courage, even in this free country of ours, to advance and support ideas that are contrary to the trend of popular, professional thought. I am proud of your nerve in publishing [the article], but I have no suggested solution to the problem discussed."[12] Melis's closing words, avoiding a personal commitment to the changes Weaver proposed, clarify the reality of actual change possible at that time within the Forest Service. There were very few supportive letters.

Harold Biswell later credited Harold Weaver's breakthrough 1943 article with elevating his own efforts to promote prescribed burning in ponderosa pine. "Weaver was roundly criticized for his views, but continued his observations of the debris-filled, disease- and insect-riddled forests, conditions that he thought were due to fire exclusion. He was clearly ahead of the times."[13] As was Biswell. The two men would, in the 1950s, become close friends and allies. They shared interests in fire ecology and prescribed burning, but also shared the experience as targets of fierce controversy and professional attacks that came with being "ahead of the times" on the subject of fire. The research findings and practical demonstrations of both men would sometimes be dismissed simply because they were "outsiders"—Biswell a research scientist, rather than an administrator of wildlands, and Weaver, the forester who dared to reopen the "light-burning" controversy in the West, was not a *U.S. Forest Service* forester.

In 1948 Weaver moved to Phoenix to become the BIA area forester. There, each autumn, working with yet another "Harold"—forest manager Harry Kallender—they burned dense ponderosa pine stands on the Fort Apache Indian Reservation. About 65,000 acres were prescription burned in November and December 1950 (the reservation had over 500,000 acres of ponderosa pines). Weaver reported initial results in a 1951 *Journal of Forestry* article.[14] Each time he wrote about the role of fire in pine forests, a few more letters of support would arrive; the doubters of the prevailing dogma were, little by little, being flushed from the woods. "I see," Emanuel Fritz wrote, "you are still pursuing your study of the relations between burning and silviculture. It is a worthwhile and excellent pursuit and I wish you well." Fritz, a professor of forestry at UC Berkeley, was an expert on the role of fire in the redwood region.[15] He continued, "In the early days of forestry we were altogether too dogmatic about fire and never inquired into the influence of fire on shaping the kind of virgin forests we inherited. Now we have to 'eat crow.' Keep up the good work."[16]

Encouragement from an academic was important to Weaver, whose articles reported the results of BIA burns primarily as practical demonstrations and always closed with a call for further formal research. He wrote back to Fritz, grateful for his support, and also corresponded that autumn with Professor H. H. Chapman, whose persistence had finally paid off in the South. "It has been my hope," Weaver told Chapman, "in the study of ecological significance of fire in ponderosa pine . . . that some real forest research organization will take it up and make a thorough study. In any event your work in longleaf pine of the south has made our path much easier, though it still is rough in places."[17]

From 1951 to 1953 the average number of wildfire acres on the Fort Apache reservation was reduced by 99 percent on lands that had been prescribe burned—one-ninth the rate of wildfires on untreated acreage elsewhere on the reservation.

Weaver's reports and articles provided the answer to an argument relied upon heavily by light-burning foes in California early in the century, that fire must inevitably damage forest reproduction by killing young trees. In a 1956 article, "Wild Fires Threaten Ponderosa Pine Forests," Weaver began by posing that very question: "How indeed have the Big Trees survived a thousand years of long, dry summer seasons with occasional dry lightning storms so characteristic of the ponderosa pine region? How do we account even for ponderosa pines 300 years old in areas where we know there have been many fires?" He concluded: "There's a reason all right. It's not that the trees were fireproof. Rather they survived so long because the forest was 'conditioned' to fire by fire itself."[18] Long-term perpetuation of the forests before modern fire prevention was possible because seedlings of ponderosa pine and giant sequoias established themselves wherever bare soil was exposed by wind-thrown tree roots overturning soil or where clearings were created when fires consumed snags or windfalls from insect- or lightning-deadened trees. Even-aged groupings of trees were the result, a characteristic of most of the ponderosa pine region.

Weaver wrote to Biswell, in 1956, about the Fort Apache burning program:

> I am glad that you like the prescribed burning in Arizona. With respect to the acreage covered, however, I believe that we have gone just about

as fast as we can. Our forestry staff is very small and we have had to combat considerable prejudice and numerous objections by critics. Mr. Ben Avery, a self-appointed high priest of conservation and a feature writer on the *Arizona Republic*, has seen fit to throw all the dirt he can on it. Unfortunately, last summer an experimental fire for control of brush set by Dr. Robert Humphrey, Professor of Range Management, University of Arizona, in cooperation with the Forest Service, escaped control and covered approximately 16,000 acres on Mingus Mountain in Yavapai County. Uncle Ben has made the most of this opportunity and has succeeded in thoroughly confusing the issue.

Our strongest advocates are the Salt River Water Users Association, the Arizona Cattlemen and the editorial staff of the *Arizona Farmer*, who have taken up the cudgels against Uncle Ben. Every time the cattlemen meet in Arizona they now give the Forest Service a bad time, asking when they're going to start burning like the Fort Apache Reservation has. As you can imagine, this has not endeared us in certain quarters.[19]

Harry Kallender (the coincidence of "Harolds" was amazing) carried on the Fort Apache reservation burn program for many years, using "men on horseback throwing matches and some on foot using kerosene drip torches," so that costs of applying the fire were kept to 2.5 cents per acre.[20]

In describing the fire history of western forests, Weaver, in those early years, tended to downplay the historic role of intentional Indian burning. While he hoped to influence the broader forestry profession, his work was done on Indian land for the Bureau of Indian Affairs, and he must have felt a particular need to buffer himself against foresters prone to simply dismiss prescribed burning as "Paiute forestry"—that derogatory label used by Chief Forester Henry Graves and others toward the light-burners in California forty years earlier. In his comments for the script of a slide show being developed by the Arizona Watershed Management Division, Weaver said, "I would leave the Indians clear out of the picture. They may have a certain selling value for some of your listeners and viewers, but for conservationists as a whole and foresters in particular prescribed burning by Indians will fall with a dull thud. For an example of what can happen to people who advocate 'Paiute Forestry' you should review Stewart Edward White's attempts to preach 'Indian or Light Burning.' " And, regarding a slide later in the program, he said "you should explain that foresters are supervising this burning—not Indians. Prescribed burning is a dangerous tool that must be expertly applied under proper conditions to keep it under control. You should stress the skill and care needed in its application."[21]

Harold Biswell was also publishing the results of his burning research and began hosting annual field days to Hoberg's Resort and Teaford Forest to show people the research on-the-ground. When weather and fuel conditions were right, he conducted demonstration burns. Communication was one of his strengths, pursued with the boundless energy he devoted to every task, not only through academic publications, but also speeches before a wide range of community and academic groups and articles in popular journals.

Controversy kept intensifying throughout the 1950s. "I was continually told that broadcast burning could not be done in California because the fuels were either too wet or too dry for burning, and the slopes too steep. The state forester told me in 1950 . . . to stay in the brush and grazing lands below and keep out of the forests. They did not want me up there in the pine forests!"[22] Fred Baker, Dean of the School of Forestry, wrote him on December 9, 1949:

> Dear Biswell,
>
> Last Saturday at the meeting of the Society of American Foresters, I had a long talk with a number of the staff of the State Division of Forestry. They are very much worried about the great increase in the amount of brush burning and the possible use of some of your tentative findings as a lever to reduce the State's appropriation for fire protection. There is apparently a rather well founded fear that this material may be used in an endeavor to prove that money for fire protection in the brush areas is not needed, since brush burning is desirable rather than harmful, and secondly that the removal of the brush will produce a far greater water crop.
>
> I think there may be a good deal of truth in their viewpoint and I want again to state that we must be exceedingly careful in the wording of every statement that is made so that it will not go an inch beyond the observed facts. Furthermore, I think that perhaps you should make even a stronger statement . . . to point out clearly that the material is not to be considered final, in addition to your present statement that it is not for publication.[23]

After Biswell's speech to the California Botanical Society on November 17, 1949, he asked the Dean to review the text of his talk and received a three-page reply ending: "As long as we are working in this highly controversial field, statements may be lifted from their context and used in a way which was never intended. I think therefore

A field demonstration of prescribed burning by Harold Biswell. Courtesy of Bruce Kilgore and the National Park Service.

that it is necessary that you should use the most extraordinary care in your statements—make sure that they hold together and are thoroughly established and are on the conservative side rather than the radical. I am sure that whatever you do there are going to be a lot of people who will take issue with you, and I want you to be ready to have an ironclad defense at all times."[24]

The dean, though worried about controversy, was at first supportive, but by 1951 pressure on Baker began mounting and he transferred it onto Biswell. A field day was held on Saturday, April 5, 1952, at Hoberg's Resort in Lake County to demonstrate work on brush manipulation and prescribed burning in second-growth ponderosa pine. Biswell's invitations to that field trip included an outline for discussions following the demonstrations and tour, covering:

The original ponderosa pine forest and the natural occurrence of fire
Working against and with nature
Prescribed burning in the southern pine forests and elsewhere

Light burning studies in California in 1920–1922

Possible reasons for prescribed burning in second growth ponderosa pine forest areas:

To reduce fuel and the hazard of wildfires

To manipulate brush

To improve vigor of trees and reduce bug damage

To obtain forest reproduction

To improve conditions for game and to facilitate hunting

To improve composition of forest stand

To ease blister rust control work

To increase water yield

To increase abundance of nitrogen-building legumes

To prevent stagnation of forest stand

To avoid accelerated erosion that might follow destructive wildfires and to prevent loss of reservoir capacity through siltation

To prevent siltation of streams and the destruction of fish that follow large wildfires

To reduce the number of pine seed-eating rodents

To improve conditions for recreation

To reduce the number of lightning fires.

Such a comprehensive list was guaranteed to rouse a concerned response from the suspicious fire-suppression community. Final topics included in the outline were "economics, techniques of prescribed burning, and the need for further study."[25]

A few days after the field trip, Dean Baker wrote a memorandum that Biswell interpreted as a threat that could cut short his career at the university. Baker described the field trip as "a remarkably fine affair in many ways." He liked the discussion there, on the ground "where the chances of misunderstanding are at a minimum" and the wide variety of points of view exchanged, adding, "I hope you continue to carry on these demonstrations." But then the tone of the letter changed:

> Although I am free to comment most favorably upon the general plan of such meetings and the way in which you handled the discussion, I feel certain that you should exercise the greatest of care in plunging into as uncertain a field as that of the use of fire in forest protection. The reasons are basically two: (1) You know very well the traditional

> viewpoint of foresters regarding the use of fire in the woods, and I am sure you realize that it is a matter in which many people take a radically different line of thought than foresters. . . . publicity may readily be spread rather widely over the region where there is a strong tendency towards incendiarism that will make it appear that the School of Forestry is advocating broadcast burning with very little of the necessary restraint and skill in the application of this work.

The Dean mentioned another member of the School of Forestry, Keith Arnold, who worked in the field of forest fire control and said, "It is far more apropos to his research work than to yours to get into it too deeply. (2) I feel that you should be more conservative in the material which you are presenting and should confine yourself to the use of fire in the improvement of range rather than in safeguarding the forest. . . . you are tackling too big and broad a subject, and your background and training are such that you are unable to do it in a highly scholarly manner."

This last was a criticism that Biswell would face repeatedly: he was an "ecologist" commenting on matters of "professional forestry." That his ecology background automatically reflected poorly on the "scholarly manner" of his research must have been a difficult criticism to swallow, particularly as it came from the dean of his own school. Baker continued:

> Although very little was said aloud at the meeting on Saturday, there was a great deal of opposition to your viewpoint expressed to me since that time from the standpoint of effective protection, fire behavior, good silviculture, and wild land management and economics. In view of this sentiment, I also think it would be very wise of you to withdraw your work from Hoberg's Resort area, since, if any group becomes sufficiently opposed to your program to try to undermine it in any way, you are leaving yourself open to very realistic criticism in doing some fine clean-up work for a well-to-do resort owner out of funds that are raised by taxation.
>
> . . . I want you to understand that I am in no way opposed to your general method of keeping every interest well informed of your work, but I wish you would be very careful to make it understood that it is your own work, and does not necessarily carry the okay of the school of Forestry as a whole. Also I would like to have you . . . make sure that your points of view are not given too much publicity.

In essence, he could do his research, but quietly, without notice, and should stop short of suggesting any alterations in forestry policy. The Dean concluded:

> You remember that when you called upon me for comments in your discussion you pointed out that I was having a good sleep there on the ground. As a matter of fact, I was not sleeping in any sense of the word. I was very much worried about the things you were saying, since I felt that I could deny the validity of every single one of them by quotations from authorities that were just as good as those you were using in your behalf. You really had me distinctly worried, and the comments I have been receiving from many sources leave me in the same frame of mind. Please use your same field meeting methods, but be extremely careful what you say and how you say it.[26]

Again the validity of his work was being slammed. How this letter must have stung! Dean Baker multiplied the pain even more, however, by sending carbon copies of the memorandum outside of the university, to the state forester, the chairman of the State Board of Forestry, and to the head of fire control at the Forest Experiment Station.

"I read this letter three or four times," Biswell later wrote, "and decided that, unless I could drum up support, I would either have to stop this particular research or be dismissed from the university. I wanted neither alternative."[27] He began drumming.

Paul Sharp, director of the University Agricultural Experiment Station sent him a letter that praised his work and encouraged him to hold more field days. Sharp also called Baker into his office to scold him for sending copies of his critical memorandum outside the university.

The Dean apologized to Biswell in writing: "There is no use in our getting too deadly serious on this affair. You are active, a hard worker, and a valuable man. Now don't let my viewpoints bother you too much. The main thing is that you have not persuaded me personally in the correctness of your viewpoints."[28]

University of California at Davis faculty Ben Madson, George Hart, and Tracy Storer also rallied behind Biswell, traveling to Berkeley to see Claude Hutchinson, dean of the College of Agriculture. They asked Hutchinson to promise never to sign dismissal papers for Biswell's release from the university.

His burning studies and field trips continued, as did his public statements about the relevance of the work to forest management. In fact, he was becoming more and more convinced that the fire exclusion policy would have to change, as shown in a letter to the State Horticultural Inspector of Oregon dated August 18, 1952. He described his research in brushlands, woodland-grass habitats, and ponderosa pine where "the burning is done in the winter—first by broadcast burning followed by piling and burning the remainder of the dead material. This works fine. On the surface it appears to cost quite a bit (piling chiefly) but *in the long run I think it will be something the forester cannot afford not to do*" [italics added].[29]

Biswell and Harold Weaver first met in 1951. They began a long relationship, reviewing each others papers before publication, commenting on each others projects and commiserating with each others' trials. After the 1952 field trip that produced Dean Baker's admonitions, Biswell told Weaver about the episode and its aftermath. "The trip caused quite a stir among foresters," he wrote. "I think probably that the work looks so good they are worried that there will be a big interest and a demand for more work of this sort." As for the institutional worries, "The final result, I think, will be a firmer basis for the work and will probably involve others for various phases of the work. Anyway, the work has not stopped!"[30]

Biswell invited Weaver to attend his 1953 field day at Hoberg's "if at all possible and . . . give us a 20-minute talk on 'Fire as an Enemy, Friend and Tool in Forest Management.' "[31] However, Weaver was working in the Washington, D.C., office of the Bureau of Indian Affairs then and could not make the trip out. In his March 1953 regrets sent to Biswell, he added, "After I get in my stint here in the Washington Office . . . I hope that there will be some possible way to get into research work revolving around my specialty. There is nothing else that I prefer doing, but the pressure of this administrative work makes it increasingly difficult to follow it. Advancement in this bureau leads farther and farther away from the woods and along paths not to my liking. I certainly hope that by another year it will be possible to visit you in California and to see your work. I will appreciate hearing how it goes at Hoberg's."[32]

One year later Weaver *was* able to attend the 1954 field day and his talk, using the title suggested a year earlier by Biswell, was billed as a "special feature" in the invitations Biswell sent out. Weaver wrote afterward to say how much he had enjoyed the day at Hoberg's: "You have done an outstanding job of cleaning up the ground and of reducing hazard. I wonder if you shouldn't give [hazard reduction] much the principal emphasis in presenting your work."[33] Weaver's advice was, of course, exactly the opposite of that Biswell had earlier received from Dean Baker.

In the same letter Weaver added that he was "extremely provoked" by the annual report of the Forest Service's Rocky Mountain Forest and Range Experiment Station relating to work done at Fort Apache reservation in Arizona.

> It is a pessimistic report and I can see no reason for it being so. We have proven that our work has had pronounced effect in reducing hazard, number of fires, area burned and cost of suppression and damage. The data is available, but they have ignored it. As for damaging the soil, I can't get much excited. In view of the fact that present forest soils have developed over thousands of years of periodic burning, why should it be damaged by properly conducted prescribed burning? I am wondering about the advisability of taking issue with the statements in the report . . . but wonder if quarreling with them now might prevent future cooperation. Perhaps they want it that way.

Weaver shared good news that he had just accepted a transfer to Portland, Oregon as BIA area forester. "I hope to be able to follow up on the prescribed burning work that we started up on the Colville Indian Reservation in north Washington and perhaps to start something new," he wrote. "Anyway, I have ideas."[34]

Rather than lay low and avoid public notice of his work, Biswell kept a high profile in both professional and public forums. He authored "Forest Tinder in Ponderosa Pine" for *California Agriculture* in October 1956. He gave as many as ten speeches in some months to campus groups, Rotary Clubs, and others. When Representative Clair Engle, chairman of the House Committee on Interior and Insular Affairs, called for a congressional investigation into U.S. Forest Service fire-fighting practices, Dr. Biswell was asked to testify before the committee.[35]

On the night of November 25, 1956, eleven fire fighters had died fighting the Inaja wildfire in the Cleveland National Forest in San Diego County. During that same fire season, large wildfires, pushed by southern California's infamous "Santa Ana winds," had burned in the San Bernardino Mountains and destroyed movie stars' homes in Malibu. Calling for hearings (that would be held in 1957), Engle claimed that Indians had managed the land better than the white men who displaced them and then bluntly declared that "the Forest Service doesn't know its business. It has permitted impenetrable and highly inflammable piles of brush to grow up just begging to catch on fire and burn up everything in the area."[36]

The *San Bernardino Daily Sun* newspaper contacted Biswell in January 1957 because they had just begun a twelve-part series (to run from January 3 through February 8), on controlled burning of brush land as a fire pre-suppression measure. Each article interviewed a proponent or opponent of controlled burning, among them a forest supervisor, the director of the State Department of Natural Resources, and the president of a local soil conservation district.

The seventh in that series quoted Albert R. Swarthout, eighty-five years old at that time, who was the only forest ranger in the San Bernardino Mountains from 1898 to 1905. "I see those forest signs 'Keep our forest green,' " Swarthout said, "and can't help thinking they should read 'Keep our forests clean'. In the old days, lightning caused most fires. . . . However, this lightning normally was accompanied by rain which held down the fire. Even so, we would let patches burn. And in the winter, also, we would clear buffer lands and do light burning in sections. We didn't let it get too heavy like now, when the brush can create fires big enough to cause extensive damage and burn timber. Fight fire with fire," he said, "and work with Mother Nature." He pulled the curtain aside to look out the window toward the fire-scarred mountains. "I hate to see all this happening. It isn't necessary."[37]

Dr. Biswell's successful burns in northern California chaparral were described in the ninth article in the series. The final article quoted James K. Mace, deputy state forester in the regional headquarters in Riverside, who said, "A fire of 50 acres of typical brushland in Southern California releases the energy equivalent to a 20-kiloton atomic bomb." Mace saw little hope at that point in controlled burning, though he called for more research to develop prescriptions. His focus was on the need for wide firebreaks, access roads, fire suppressant chemicals, and aerial attack technology. "Let's not forget that it only takes one little unnoticed spark drifting across a control line or firebreak to turn a smoothly operating control burn into a holocaust."

After the series finished, Biswell wrote their author, Donald R. Geggie, saying: "There seems to be quite a difference of opinion about what should be done. I think this means that research and demonstration needs to be done to learn more about it. I think perhaps the solution is to select watersheds of 10,000 acres or so and test some of these ideas. I expect, however, that the whole thing will be forgotten about soon and sometime in the future we will have a fire worse than any of those last summer."[38]

Biswell participated in Engle's congressional committee hearing in October 1957. His statement emphasized his regular themes. Fire suppression, however essential around development and forest-use areas, was "*fighting nature* and such action always results in trouble," he told the congressmen. The choice to build more access roads and intensify suppression efforts "will not solve the fire problem and it intensifies soil erosion. The second choice is to reduce fuels" by prescribed burning. "In this work we are *cooperating with nature*" [italics in original]. Biswell was ready to push for practical measures and suggested that the United States "borrow an idea from the Swedish Forest Service. This Service employs professional broadcast burners. They are the elite among the Forest Service personnel, and are looked upon with the greatest respect."[39] He encouraged the state and federal services to train such a corps of specialists.[40]

The congressional investigations did pave the way for political and financial support both for expanded research and fire suppression funding, but even worse fires would be in southern California's future, as Biswell predicted, and no commitment to a corps of elite broadcast burners was forthcoming.

Weaver, at a comfortable distance from the southern California fire problem, revealed an interesting take on those issues when he wrote his entomologist friend, Paul Keen, who had moved to California from Oregon:

> I have been hearing quite a bit about the fires in California. As you say, the fire boys of California really have a tough problem to handle. I never want to become involved in it. I remember during the war when Harry Kallander was stationed in San Diego County, he once wrote that a plane crashed nearby and started a brush fire in the steep mountains. He ended by saying, "I did nothing about it." That would be my attitude. I never would build a home in some of the extremely hazardous locations that appear to attract the rich Californians. That would be asking for it to build a home in the midst of such a tinderbox.[41]

Harold Biswell stayed directly involved in all of the state's wildland fire debates. He spoke to the Society of American Foresters at their meeting in New York that year on "The Use of Fire in California Chaparral for Game Habitat Improvement." He described burning techniques and results in "true" (or climax) chaparral, woodland-grass chaparral, and in timberland chaparral associated with ponderosa pine forests.

Meanwhile, Biswell had also been following the progress of prescribed burning in the South, hoping that the breakthrough there would help his and Weaver's efforts in the West. After Elwood Demmon, the director of the Southeastern Forest Experiment Station, visited Berkeley, Biswell wrote Weaver: "I asked him how prescribed burning is coming along in the South. . . . He said that 'excellent progress is being made. It is the cheapest and most useful tool we have; its value has been proven beyond doubt. The severe fires of the past year proved to us that if we expect to sustain forestry we must prescribe burn, and we must do more than we are doing at present.' I asked Demmon about this because I was beginning to think that perhaps they were doing less burning than when I was there from 1940 to 1947. However, they are doing more. Thought you would be interested."[42]

The South's growing acceptance of prescribed fire presented an obvious opportunity to show parallel conditions that applied out west. Before submitting "Prescribed Burning in Georgia and California Compared" for publication, Biswell had the manuscript reviewed by Arthur W. Hartman, the head of U.S. Forest Service fire control in the southern region. Hartman made a number of detailed suggestions to broaden the scope of the paper's examples and strengthen the overall tone, explaining: "The reason some of my comments might appear to be a little blunt is that I feel that our profession is still dragging its feet and permitting sentiment and emotion to rationalize hesitancy to tackle and find out just what the scientific place of prescribed fire is in our various situations and problems. Jarring loose those high resistance folks with their fixed preconceived ideas is tough enough even when you use all of your ammunition."[43]

Hartman became another of Biswell's sympathetic professional correspondents. Like Weaver and Biswell, Hartman was one of the "torchbearers" then leading the profession back toward acceptance of fire as a tool. In a 1956 letter he passed along some criticism from A. A. Brown, the director of Forest Service research in the Washington, D.C., office. Brown called Biswell "headstrong and very much an extremist." Biswell could not remember Brown ever attending any of his field days or discussing fire issues directly with him, and concluded that Brown just "was talking from hearsay and imagination."[44]

The Georgia/California article was published in 1958 in the *Journal of Range Management*, after first being rejected by the *Journal of Forestry*. It reviewed the historic light-burning controversy in California, including the California Forestry Committee's conclusions of 1923. Biswell noted, mildly, that "great changes have taken place since then and some of the reasons advanced at that time for not burning are no longer valid, especially in the light of the information gained in the past 7 years."[45]

Comparing the forest ecosystems of the two regions, he asked, "If [controlled burning] is used successfully over such wide areas and under such variable conditions as those found in the South and the Southeastern states, why could it not also be used in California? This idea was also developed from the fact that frequent light fires had probably been the most important force in molding the virgin forests of California."[46]

The secretary-manager of "Keep California Green" (KCG) sent a memorandum to his board of directors that stirred up an exchange of letters in October 1958. The organization existed to "Prevent Man-Caused Wild Fires," and its directors were from timber, ranching and other corporations along with an advisory council with representatives from the U.S. Forest Service, Bureau of Land Management, California Division of Forestry; plus Dean Henry Vaux of the School of Forestry. Wayne Hubbard's memo, headed "Comments and Questions on Biswell's Fire Prevention Article," actually was reaction to a press release covering a talk Biswell gave before Ecological Society meetings in Indiana. Hubbard's memo went to Vaux and faculty member Emanuel Fritz at the U.C. Berkeley School of Forestry.

Biswell wrote to Hubbard that he was "sorry that the press release caused you some alarm. I want to make it perfectly clear to you and everyone else that I am in no way an 'opponent' of fire prevention, neither do I have a 'let 'er burn' theory. Please understand that I am just as interested in keeping California green as you and all of your members." As he invariably did, after explaining his research and enclosing reprints of published articles, Biswell invited Hubbard to "spend a day with me in the field on the experimental area where we can look and discuss more thoroughly the objectives, methods, and results. Then next spring, in late May or in early June, I should like to show your members over the experimental areas. I hope you will quote this letter verbatim for your members. *The research is too important to quarrel about!* My thought is that misunderstanding leads to controversy and can only retard progress, while knowledge and understanding can greatly accelerate it"[47] [italics added].

Emanuel Fritz provided Biswell with a copy of *his* reply to Hubbard supporting Biswell's research and emphasizing that Biswell was not an opponent of fire prevention. Fritz pointed out that "the magnificent Sierra Nevada pine forests were the result of frequent burning. Certainly they did some damage, but they also made it possible for the trees to reach great size, quality, and value.

"I have fears that the fine sequoias in areas like the Sugar Bowl on Redwood Mountain in Tulare County are going to show decadence if the competing dense white fir stands are not thinned out. There just isn't enough water to satisfy the needs of all these trees—old and new." Fritz *was* concerned about general public knowledge, however, of the sort that Biswell fostered with his speeches and articles. "These things are best kept within the small circle of progressive owners and managers," he wrote. "It is not good policy, under present conditions, to let the public get the idea that anyone can burn." And he closed with a statement that illustrates the difficulty even "progressive managers" were having with the subject: "I personally hope we will find that we can do the silvicultural job without fire."

His handwritten postscript at the bottom of Biswell's copy read: "Harold—For your own protection and to prevent the general public from getting wrong ideas, I personally hope you will keep the results of your experiments within professional journals and out of the general press for the present."[48]

Hubbard responded to replies from Biswell, Fritz, and also from Vaux, graciously emphasizing to the dean that, "I am no more an opponent of controlled burning than Dr. Biswell is of *Keep California Green*." To Fritz he wrote, "It was not my intention to openly and publicly criticize Dr. Biswell." Fritz's suggestion that the research results be kept within a small circle sounded good to Hubbard. "*Controlled* burning will not do KCG any harm, in fact, it could do fire prevention some good *if* handled by experts. My criticism has been with the impressions the public must have gotten. Beyond that I cannot criticize very much." To Biswell himself, Hubbard wrote:

> After reading the speech I find it less inflammatory than the release covering it. I have received many letters regarding your speech from people in forest land management positions. None categorically condemn controlled burning but all are concerned with the publicity given such statements as: "With the modern policy of fire suppression, the forests gradually build up a virtually solid layer of fuel from the tops of trees to the brush and young trees and dead material beneath. Under these conditions, wildfires in the summer are destroying a wealth of resources."
>
> Obviously this . . . leads to the assumption that we are "losing by

> preventing fire." If it were stated that protection agencies are not willing to use controlled burning as a protection tool, wouldn't that more clearly present your feeling? If the public is asked to disagree with fire suppression then they are asked to disagree with Keep California Green and our campaign is seriously affected.[49]

Fire prevention campaigns would always remain necessary, but Biswell saw a realistic need to counter a half century of fire exclusion dogma before a program returning fire to ecosystems had any hope of being implemented. The most successful fire prevention effort ever had been born during World War II—Smokey Bear was, by the late 1950s, a popular cultural icon. But was his success damaging prospects for peaceful coexistence with nature's fire?

NOTES

HB—Harold Biswell papers, held at the Bancroft Library, University of California, Berkeley (BANC MSS 2002/67 c).

HW—Harold Weaver collection at the Forest History Society, Durham, North Carolina.

1. April 8, 1947, memorandum from Walter Mulford, Chairman, Department of Forestry, University of California at Berkeley, to Vice President Hutchison, HB.
2. Biswell to Dr. I. T. Haig, February 18, 1947, resignation letter, HB.
3. Harold Biswell, *Prescribed Burning in California Wildlands Vegetation Management* (1989; reprint, Berkeley: University of California Press, 1999), 100.
4. Biswell, *Prescribed Burning*, 101.
5. Griggs to Biswell, October 25, 1949, HB.
6. Harold Biswell, "Fire Ecology: Past, Present, and Future," Keynote talk to the Ecology Section, American Association for the Advancement of Science, Davis, California, June 23, 1980, Speech copy, p. 5, 6, HB.
7. Harold Weaver, "Fire as an Ecological and Silvicultural Factor in the Ponderosa-Pine Region of the Pacific Slope," *Journal of Forestry* 41 (January 1943): 13.
8. Weaver to Dr. Henry Schmitz, May 29, 1942, HW, Box 1, File 1.
9. F. P. Keen to Duncan Dunning, October 1, 1942, HW, Box 1, File 1; also note letter from A. A. Brown to Harold Weaver, September 23, 1942, in same file.
10. Keen to Weaver, February 8, 1943, HW, Box 1, File 1.
11. F. C. Craighead to F. P. Keen, memorandum, February 23, 1943, HW, Box 1, File 1.

12. Percy E. Melis to Weaver, March 4, 1943, HW, Box 1, File 1.
13. Biswell, *Prescribed Burning*, 101.
14. Harold Weaver, "Fire as an Ecological Factor in the Southwestern Ponderosa Pine Forests," *Journal of Forestry* 49 (February 1951): 93–98.
15. See Emanuel Fritz, "The Role of Fire in the Redwood Region," *Journal of Forestry* 29 (October 1932): 939–950.
16. Fritz to Weaver, April 4, 1951, HW, Box 1, File 8.
17. Weaver to Chapman, October 29, 1951, HW, Box 1, File 8.
18. Harold Weaver, "Wild Fires Threaten Ponderosa Pine Forests," *American Forests* (February 1956): 28.
19. Weaver to Biswell, September 17, 1956, HB.
20. Harry R. Kallender to Biswell, March 26, 1955, HB.
21. Letter from Weaver to Mr. Joseph F. Arnold, Director, Watershed Management Division, Arizona Office of State Land Department, May 31, 1957, HB.
22. Biswell, *Prescribed Burning*, 103.
23. Baker to Biswell, December 9, 1949, HB.
24. A second letter from Baker to Biswell, also December 9, 1949, HB.
25. "Field Trip" invitation that also said "Bring your lunch; a hot dog if you wish to roast it over a bonfire in the forest," HB.
26. F. S. Baker to Dr. H. H. Biswell, April 9, 1952, memorandum, HB.
27. Biswell, *Prescribed Burning*, 105
28. Ibid, 106.
29. Biswell to Mr. J. R. Roaf, State Horticultural Inspector, Department of Agriculture, Salem, Oregon, dated August 18, 1952, HB.
30. Biswell to Weaver, May 20, 1952, HW, Box 1, File 8.
31. Biswell to Weaver, December 29, 1952, HW, Box 1, File 10. Weaver later used Biswell's suggested title for his July 1955 article for the *Journal of Forestry*.
32. Weaver to Biswell, March 24, 1953, HB.
33. Weaver to Biswell, May 30, 1954, HB.
34. Ibid.
35. Letter from Clair Engle, Chairman of Committee on Interior and Insular Affairs, to Biswell, September 23, 1957, with Biswell's note added by hand, "Pleased to accept your invitation to attend subcommittee hearings in Los Angeles on Oct. 8 & 9."; telegram acknowledged by Harris Collingwood, Forestry Consultant on September 30, HB.
36. *San Bernardino Daily Sun*, "Training Fire Fighters on Career Basis Advocated by Veteran Forest Official," January 3, 1957, by Bob Geggie. Eleven additional articles followed in the series, ending February 8.
37. *San Bernardino Daily Sun*, "Retired Ranger Favors Controlled Burning," January 21, 1957, by Bob Geggie, seventh article in the series.
38. Biswell letter to Donald R. Geggie, *San Bernardino Daily Sun*, February 28, 1957, HB.

39. Biswell cited Iggesund's (Sweden's) Forest Industries, "From Forest to Factory," Sweden, 1954.

40. H. H. Biswell, Statement before Representative Clair Engle's Subcommittee (of Committee on Interior and Insular Affairs) hearings on forest and brush fires, in Los Angeles, October 8 and 9, 1957; copy of Biswell's paper quoted here, pp. 4, 5, HB.

41. Weaver to Keen, April 10, 1957, HW, Box 1, File 11.

42. Biswell to Weaver, November 15, 1955, HW, Box 1, File 10.

43. Hartman letter to Biswell, undated, HB.

44. Biswell, *Prescribed Burning*, 107.

45. Harold Biswell, "Prescribed Burning in Georgia and California Compared," *Journal of Range Management* 11, no. 6 (November 1958): 294.

46. Ibid, 294.

47. Biswell to Hubbard, October 6, 1958, HB.

48. Fritz to Hubbard, October 9, 1958, HB.

49. Hubbard to Biswell, October 17, 1958, HB.

4. Only You

> The United States Forest Service is no advocate of the "isms" of Europe, but it is teaching a doctrine of hate in the [Mississippi] schools. . . . Here it is an American hate, not directed against a people but . . . a hate against forest fires.
>
> R. M. Conarro, 1939

> We know he's a single-minded, but positive fellow. . . . We trust Smokey; he has won our confidence.
>
> Jim Felton, former national coordinator, Smokey Bear campaign

Fire prevention campaigns were the propaganda arm of the war against "evil fire." As with traditional wartime propaganda, demonizing the enemy was often the most effective way to motivate the public. Patriotism and fire prevention became natural allies during World War II. In 1942 a Japanese submarine surfaced off the coast of southern California and fired some shells that came uncomfortably close to Los Padres National Forest. Later that year there were three attempts to fire-bomb the coastal forests of Oregon. Concerned about the threat of widespread fires that might result from incendiary attacks, the Forest Service organized a defense campaign, the Cooperative Forest Fire Prevention Program, to protect the nation's forests as a vital wartime resource.

For that campaign, the Wartime Advertising Council—advertising agencies that donated services to the war effort—produced such slo-

TOKYO Loves an American FOREST FIRE!

THE *Axis* invented the phrase *"Fifth Column"* . . . it means any kind of sabotage. It strikes at morale, property, or resources.

Forest fires, easy to start, hard to stop, and inestimably destructive in certain seasons, are a natural temptation to any enemy fifth columnist.

Too much timber is already burned by loyal but careless Americans. Let's cut this kind of conflagration.

Forest Fire is Forest Enemy Number One. Do your part to keep up the parade of new forest growth already started by the private forest industries.

World War II fire prevention message.

gans as, "TOKYO Loves an American FOREST FIRE!" and "Careless Matches Aid the Axis!" Some fire prevention posters used leering images of Hitler, the Japanese emperor, or evil-looking Nazi soldiers, gleeful about wild flames consuming American forests. Former Chief Forester William Greeley explained the focus of the program and the mindset of the aroused populace in the foreword to a history of American forest fires (Stewart Holbrook's *Burning an Empire*) published in

1943: "Today, on the Pacific Coast, we witness . . . the story of preparation and defense . . . keeping our woods from burning and our air clear from smoke—as a measure of national defense. The sheriff who declared that he would shoot any firebug on sight and try him afterward, spoke the mind of the people."[1]

"Keeping our woods from burning" had vital strategic importance. Wood had multiple uses that only in later wars would be superseded by metal and plastics. Rifle stocks, of course, were made from wood, but its utility stretched much farther. Douglas fir trees produced planks used to cross rivers on pontoon bridges, provided rails for field stretchers and the material for huts that housed soldiers. "Every man in the service knew [Douglas fir lumber] well, for his foot locker was generally made of it. Fir went into the tanks for gasoline storage at advanced bases. Plywood . . . proved ideal, too, for the Navy's lifeboats, saving a ton of steel on each boat, and floating 7 inches higher out of water than its steel counterpart. Plywood also [went] into coastal patrol and torpedo boats, into PT and assault and landing boats."[2] Even the helicopters of that era, underpowered experimental aircraft used in the Pacific theater, were kept lightweight with wooden frames (covered by canvas).[3]

Beyond wood products, preventing major forest fires also meant that men needed in the fights overseas would not have to be kept home to fight domestic forest fires.

The Japanese clearly recognized the strategic importance of America's vast western forests. Their military launched over 9,000 "fire bomb" balloons during the winter of 1944–1945, to ride the jetstream winds toward America's West Coast. Most carried five bombs—a thirty-three-pound fragmentation bomb; four, eleven-pound incendiary bombs; and demolition charges to destroy the balloon itself. Several hundred actually reached North America, "falling, or being seen in 26 states and provinces from Mexico to Alaska, and as far east as Kansas, Iowa, and Michigan."[4] But it was winter and too wet for the incendiary bombs to produce any more than locally charred circles. The U.S. military kept the invasion secret until five children and their Sunday school teacher were killed when a bomb detonated on a balloon they discovered near Bly, Oregon, on May 5, 1945.

War's particular requirements reignited debate over fire as a tool in the South. Wheeler and Chapman produced another exchange of letters in the *Journal of Forestry* in June 1944. Chapman's research now showed that fire was needed as a tool, not only in longleaf pine forests, but for loblolly and slash pine production. Wheeler remained steadfast, expressing amazement that, Chapman, "one of the most ardent controlled-burning enthusiasts . . . now advocates burning loblolly pine." Not surprisingly, Wheeler added the war argument to his general condemnation of fire: "Smoke from controlled fires interferes with the training of airplane pilots just as it does from uncontrolled fires. Therefore, no matter what our beliefs about woods burning are, is it not advisable to stop all burning until the war is over?"[5]

Wheeler's emotional commitment to fire exclusion allowed no room for "fire as a tool" heresy. He was a true believer and committed soldier, after all, in the propaganda arm of a war that by 1944 had a lengthening history of victories and bitter defeats shaping dogma and solidifying beliefs.

FIRE PREVENTION CAMPAIGNS BEFORE SMOKEY

Since 1911 the U.S. Forest Service had administered federal funds for cooperative fire prevention and control with the states. *American Forestry* devoted its entire November 1913 issue to state fire prevention reports. In California the State Board of Forestry—still battling light-burning advocates—produced a poster image of burnt stumps on a barren hillside with text that spoke to the recalcitrants: "Mr. Citizen—Mr. Official—Mr. Timberman—Will Your Children Have Timber to Burn?"

Pennsylvania's poster proclaimed forest fires to be "A Curse to the People of Pennsylvania. Which Would You Rather Have?

forest fires		green forests
floods		pure water
disease	OR	health
destruction		thriving industries
devastation		prosperity"[6]

While many fire prevention programs kept to the fundamental message that basic carelessness produced destructive wildfires, fire was regularly demonized as pure evil. The Grim Reaper torched the woods on posters captioned "Death Rides the Forest." Other posters and billboards depicted flaming devils with slogans "Stop Demon Fire" or "The Destroyer—Keep Him Out of the California Woods."

Fire prevention posters that demonized fire.

As Americans watched the start of World War II, the southern fire prevention crusaders in Mississippi extolled a form of righteous hatred: "The United States Forest Service is no advocate of the 'isms' of Europe, but it is teaching a doctrine of hate in the schools in the national forests in Mississippi. Here it is an American hate, not directed against a people but against the needless, ruthless destruction of our most valuable natural resources. Specifically, it is a hate against forest fires."[7]

That article, "Fighting Tomorrow's Fires Today," appeared in a 1939 edition of *American Forests* that was, again, devoted entirely to fire prevention. The editor's essay prefacing the articles was a martial call:

> Forest fires, once started are fought like wars—on the firing line. And as with wars, forest fires are prevented by a vibrant public will to prevent them. American needs that will—terribly. This issue of AMERICAN FORESTS is dedicated to the creation of such a will. America

> needs a rebellion of consciousness to bring to life a public will to stop forest fires . . . to bring to bay with guns of public opinion a national enemy that is dropping fire brands somewhere in our land at the rate of one every three minutes day and night. After reading, pass your copy to a friend or neighbor, to a library or school. Make it a circulating hammer for the forging of that public will to prevent and control forest fires without which there can be no sure or lasting preservation of those natural heritages that sustain American life.
>
> Patriotism in America may lack the virility of old and with some it may be out of fashion, but it is not dead. There are still millions of people who . . . are ready and eager to become Paul Reveres on the public opinion highways of our country. By those highways only, can we ever hope to attain a public will to prevent forest fires.[8]

The magazine contained the articles "Fire or Forestry—The South's Great Problem," by a congressional representative from Mississippi; "Burning Wildlife—20,000,000 Acres of Wildlife Domain Are Burned Annually," by the chief of the U.S. Bureau of Biological Survey; and "California Pays the Red Piper," by the conservation chairman of the State Chamber of Commerce. The director of the National Park Service authored "Outdoor Recreation—Gone With the Flames," illustrated with a park service sign: "FIRE! Keep it away from our National Parks." The large letters of "fire" were formed by flames.

A fire prevention document that tested the line separating church and state was "*Forest and Flame in the Bible*," developed for Wheeler's spring lecture tour through the South in 1937. It was an official program aid of the Cooperative Forest Fire Prevention Campaign, (still in print as recently as 1961). Selected biblical passages were chosen "and approved for this purpose by religious leaders representing the Protestant, Catholic, and Jewish faiths. They contain inspiration for discussion in our Sunday Schools, Bible Classes, and homes."[9] A series of "lesson statements" were followed by illustrative Bible passages, such as: "And keep fire from the trees, however small the flame. *James 3:5 '. . . Behold, how great a matter a little fire kindleth!' Psalms 83:14 'As the fire burneth a wood, and as the flame setteth the mountains on fire;'* For fire can destroy all that grows. *Joel 2:3 "A fire devoureth before them; and behind them a flame burneth: . . . behind them a desolate wilderness; yea, and nothing shall escape them'* "[10] [italics in original]. On the back cover of the 1961 reprint, a bear was pictured kneeling while deer, squirrels and foxes, quail and other birds watched. Wheeler's 1937 version did not include the bear's prayer—". . . and please make people careful, amen"—nor the closing slogan: "Remember—Only *You* Can Prevent Forest Fires!" Smokey Bear was not "born" until 1944.

"PROWLIN' AND A GROWLIN' AND A SNIFFIN' THE AIR"[11]

While World War II was raging, Walt Disney's movie "Bambi" was released to theaters. The 1943 movie's forest fire scene, with animals fleeing in terror, formed the dramatic climax of the animated feature. Though Bambi and his father, along with Thumper the rabbit and Flower the skunk, did survive, and the story concluded the following spring in a forest full of beauty and life, a fire ecology lesson was *not* what movie-goers took away. Rather, Bambi's danger and panicked flight made him a natural as a symbol, particularly as fire prevention campaigns were directed more at children. Bambi's role as a fire-prevention deer did not last, however. There were copyright issues with Disney that made his long-term, widespread use difficult.

So the Wartime Advertising Council considered various alternative animals, including squirrels and birds. They settled on a bear—a symbol of strength, capable of growling his message when sternness was called for, but best of all, a bear could "naturally" stand on his hind legs, wear a ranger's campaign hat, and carry a shovel. With anthropomorphic features the cartoon bear would become a cultural icon.

The first description of Smokey Bear given to artists for the Cooperative Forest Fire Prevention Program directed that his fur be black or brown; his face should have a short nose, like a panda's, and have a knowledgeable, appealing expression; and he should wear a campaign or Boy Scout hat that typified the outdoors and the woods. At first Smokey did not wear pants; blue jeans were added, soon, with a belt and buckle, the name "Smokey" appeared on his hat, and he sometimes carried a shovel.[12]

August 9, 1944, was Smokey Bear's "birthday." His birthplace was the Foote, Cone and Belding advertising agency in Los Angeles, which handled the Smokey Bear fire prevention campaign for the Ad Council (and still does today). The first poster showed the bear with a small hat, pouring water on an abandoned campfire. The caption was "Smokey Says, Care *Will* Prevent 9 Out of 10 Woods Fires!"

After World War II Smokey's artists drew him taller and friendlier. Rudy Wendelin was the artist who first drew his front paws to look like hands, which enabled Smokey to do even more human tasks.

Wendelin, a Forest Service headquarters employee, became known as "Smokey's artist," though he did not do the art for the annual Advertising Council campaigns. But when the Forest Service needed Smokey's image on special posters, educational material and for special events, Wendelin was his artist. Smokey would gradually evolve from "bear" toward a wise and fatherly character, stern, where necessary, Nature's friend, but, above all, the foe of wildfire.

Smokey, the fire prevention bear, found lasting success where other fire prevention images slipped away. Ugly faces of the wartime enemy never went over well with children in particular. Plans for one alternative symbol were dropped in 1952 because of Smokey's popularity. "Johnny, the bellhop," was then the commercial character used in Philip Morris tobacco company advertisements. The American Forest Products Industries, a trade association, had planned for the bellhop to visit tree farms (for whatever reason) where he would be very careful with his cigarettes. Fire prevention campaigns depended upon good public relations and financial support from industry, but Johnny, promoter of cigarettes, fortunately stood no chance against warm, fuzzy Smokey Bear.[13]

The famous message, "Remember, Only You Can Prevent Forest Fires" appeared in 1947 for the first time. In one of the earliest posters that used the phrase, Smokey pointed to a burning forest full of flames and smoke while he cradled a fawn in his other arm. The poster read: "Our Most Shameful WASTE! Remember—Only *You* Can PREVENT FOREST FIRES!

Focusing on helpless animals, particularly baby animals, was an effective advertising theme which partially explains the incident that began on a May morning in 1950. The Capitan Gap fire had been fought for two days in the Lincoln National Forest in New Mexico. A fire crew found a small bear cub up a tree with mild burns on his paws and rear. He was taken by Game warden Ray Bell to a veterinarian in Santa Fe for treatment. Bell later took the cub home and nursed him. His four-year-old daughter, Judy, and the family's dog, a black cocker spaniel named Jet, became the bear's friends, though the cub was grouchy and tended to make surprise attacks on Bell. The cub apparently associated him with the trauma that had brought them together.

News stories around the nation carried the story. The "Smokey Bear" name and image had been established for six years by then as the national fire prevention symbol. So the cub was named "Smokey" and sent off to the National Zoo in Washington, D.C., to be the "living symbol" of the fire prevention bear. Smokey remained at the zoo until his death in 1976. He was buried in Capitan, New Mexico, in Smokey Bear State Park. After his death a replacement was sent to the zoo.

When "Little Smokey" died in 1990, no replacement was chosen. The real Smokey was actually considered a problem by most Smokey Bear campaign staffers. A real live bear did not fit the cartoon image or character that had been carefully crafted for Smokey, so the living animal was mostly ignored by the advertising staff at Foote, Cone and Belding. A preference for unreality, some would later contend, was a problematic symptom of Smokey's message, when it sometimes came in conflict with facts about fire ecology.

Every Smokey Bear stuffed animal made by Ideal Toys in 1952 included an application to become a Junior Forest Ranger. The application was to be mailed to Smokey Bear Headquarters. Three years later there were a half-million Junior Forest Rangers. Over 5 million eventually joined. Smokey Bear received so much mail, he was given his own zip code. His official address is Smokey Bear Headquarters, Washington, D.C., 20252.

"Smokey the Bear," the song, came out in 1952, with a "the" inserted into his name by the songwriters.[14] Administrators of the Smokey Bear program have, ever since, chosen to fight a quixotic, futile battle when they send educational materials to schools and libraries, trying to overcome the widespread mispronunciation of Smokey Bear's true name.

Smokey's cadre of professional imagemakers knew their business, however. Early surveys showed that only Santa Claus and Mickey Mouse were as well recognized as cultural icons. A market research survey in 1976 found that 98 percent of those surveyed had heard of Smokey. Donald Hansen, then manager of the Forest Service's Smokey Bear campaign, said "We feel we got beat out by Santa Claus and Ronald MacDonald, but that's all."[15]

Smokey Bear's overriding mission was to prevent unwanted, human-caused fires. Toward that end, each year the campaign would focus on new slogans and themes. Some have been:

"Prevent Forest Fires!"

"Thanks for Listening."

"Smokey's Friends Don't Play With Matches. Only *you* can prevent forest fires."

"The Blazing Forest." Courtesy of the Smokey Bear fire prevention campaign, USDA Forest Service.

"Remember—only *you* can PREVENT FOREST FIRES!"

"PLEASE! Only *you* can prevent forest fires"

In 1974, for his thirtieth "birthday," the one-word message was: "Think." In a series of posters, the letter "i" within "think"was replaced by a match, a tree, a little Smokey with his shovel, or a fawn.

By 1985 Smokey's most popular refrain was so well known that only two words were required to get across his message. Smokey, in an Uncle Sam war recruiting posture, pointed at the reader, saying, simply: "ONLY YOU."

"Green" nature appreciation and conservation themes occasionally mixed with Smokey's standard "red" messages of strict fire prevention. Smokey is wise and caring and values trees and entire forests full of wildlife. In the 1980s a popular series of educational posters with beautifully detailed art depicted assortments of birds, bird nests, leaves, tree types, mammals, amphibians, insects, reptiles and other thematic categories, all presented by Smokey, but as a diminutive image somewhere on the posters. Those conservation themes were sometimes seen among program administrators as straying too far from, and potentially diluting, Smokey's primary message of fire prevention.

More in keeping with his mission was one poster that read: "One match. That's all it takes to destroy a forest and every creature, great and small, who lives there."[16] That was the kind of inaccurate message that disturbed fire ecologists.

How much influence has Smokey Bear had? Has he convinced several generations of malleable school children that fire is always bad and should have no place in the forests (even if that specific message was not exactly what his handlers intended)? Consider the pervasiveness of the campaign and its appeal to children. Tucked away among the artifacts of my youth is a "Smokey Bear Reading Club" certificate, earned by reading ten library books and "taking the Conservation Pledge." It has the signature of our librarian, but far better, it was also "signed" with Smokey's pawprint. A picture depicts Smokey showing baby trees to two baby bear cubs, beside a sign: "Smokey says: Be careful—BECAUSE EVEN LITTLE FIRES KILL LITTLE TREES." This was the gentle, impassioned, nature-loving bear I grew up with.

There, on that certificate, tailored to children, was the foresters' argument for fire exclusion—fire and forest reproduction were incompatible. Was Smokey's simple message, then, too simple? Or did that simplification matter, so long as fire safety was the overall objective?

Debates about Smokey's focus on fire prevention would grow in the final decades of the century, as the fire exclusion policy itself was increasingly questioned. Discussions about Smokey typically took on a surreal nature, sounding as if he was very real. His character became so "alive" for people, that one could plausibly rationalize Smokey's obsessive compulsion, tracing it to the life-shaping trauma of that forest fire he endured as a cub. One of the Smokey Bear program's former national coordinators, Jim Felton, once said, "Smokey Bear is a real personality with a real identity to which

"Smokey Says: Be careful—because even little fires kill little trees!" Courtesy of the Smokey Bear fire prevention campaign, USDA Forest Service.

millions of Americans closely relate. People know and understand Smokey. He is not a stranger. We know he's a single-minded but positive fellow, so we accept immediately his admonishments as the caring direction of an old friend. We trust Smokey; he has won our confidence."[17]

If Smokey Bear were as amenable to reason as his character development suggested, it would have been interesting to take the real bear aside for some re-education about what led to that 1950 holocaust in New Mexico that so changed his life. Some careless person started that fire, but its devastation was a product of decades of fire exclusion in an ecosystem where the forest, including its black bears, had evolved and adapted to frequent, low intensity fires.

In the 1950s, evidence supporting such ecological understanding was just starting to be developed by researchers. Two decades later, as fire management policies were finally changing, the media would seldom resist making a newsworthy connection with Smokey Bear. *Time* magazine led an August 7, 1972, article, "The Fires Next Time," with a quote from national park biologist Lloyd Loope, "Smokey the Bear has been lying for years." Dean Wohlgemuth asked, in the title of an article for *Georgia Game & Fish*, "Do Forest Managers Have a Smokey Bear Complex?" "Has Smokey done too well, perhaps?" Wohlgemuth added. "He just might have oversold his product. But don't blame Smokey, it's really not his fault. He was given a job to do, and he did it admirably. His job was to prevent wildfire, *not* to stop everyone from using fire for good purposes. It's just that he did his job so well that folks just won't touch a match to the woods when they need to."[18] On October 20, 1972, a front page headline in the *Wall Street Journal* read, "For Shame, Smokey! Why Are You Setting That Forest Fire?: Actually, Rangers & Ecologists Now Say, an Occasional Fire May Do Animals, Trees, Good."

One of Smokey's biographers concluded his book with this thought: "Smokey Bear lives in the hearts of the nation's children. In their minds, Smokey is the unchallenged symbol of truth, bravery and commitment, and kids have cultivated a loving passion—perhaps even a reverence—for that furry American legend."[19] Of course, WE are those grown-up children who revere Smokey Bear. Might it be we have been brainwashed to a point where propaganda diminishes prospects for any transition to peaceful coexistence with nature's fires?

Biology professor Richard Vogl made such a case. "Since forest fires are emotionally condemned as evil threats to our society, their suppression is unanimously supported," Vogl wrote in 1973,

> almost as automatically as Congress voted funds for Vietnam during the early days of the war. Who could believe, as we sat securely on mother's lap and listened to the story of Bambi, that we were being pumped with a great deal of nonsense? Who could mistrust the honest-looking, clean-cut man wearing a green uniform and badge who visited our fifth-grade class and told us about Smokey the Bear and the destruction wrought by forest fires (a program that is still fully underway with all of its misinformation)? Was he pushing a line of half-truths because he believed it intrinsically, or did he assume that we were too naïve to handle the whole truth?[20]

Ed Komarek, who through the Tall Timbers Research Station would found an influential series of fire ecology conferences in the 1960s, was quoted in the 1972 *Wall Street Journal* article: "I was educated against fire and actually fought it. Then I realized I was dumb as a coot. Fire is a natural thing, like wind and rain, and should be allowed to run its course."

As that era of change arrived, Forest Service researcher Bob Mutch became concerned about the public relations disaster that a "Smokey Bear backlash" could generate. He encouraged fire managers to urge media to resist the impulse to open every story with Smokey as an attention-getter. Fire prevention and wise fire use, after all, were not incompatible.[21]

Harold Biswell found the Smokey Bear phenomenon less frustrating than some fire researchers. It intrigued him, evidently, as a study of human psychology. Biswell said, "One of the most interesting aspects of this research has been observations on people, mainly professional land managers and others interested in environmental conservation and preservation. We must realize that people and their politics are an important ecological and sociological component of any wildland ecosystem." A German silviculturist had told him, "In California, with its dry summers and high danger of wildfires, I think prescribed burning might be a good thing. However, I wouldn't be able to do it myself. My early training led me to believe that all forest fires are bad and when I see charcoal in a forest it simply rubs me the wrong way."

"Perhaps," Biswell said, "this would indicate that we should be giving more emphasis to fire ecology in our teaching and not quite so much to 'Smokey the Bear' syndrome." But he was not too taken with the argument "that the general public has listened to Smokey the Bear so long that it will be difficult for them to see any beneficial effects of using fire as a tool. I have found that the general public grasps the idea very quickly and often asks, 'Why didn't you start it long ago?' "[22]

World War II not only reinforced the patriotism theme in fire prevention campaigns and led to the introduction of Smokey Bear, but after the war fire suppression agencies received a major boost from military technology, particularly aircraft. Parachute drops, aerial reconnaissance, and direct attack with retardants and water-drops arrived with modern planes and helicopters. Fire fighters had better tools to bolster their belief that victory over the nation's other great "enemy," wildland fire, was truly achievable. For awhile they would enjoy more success. Then, increasingly large losses, as fuels kept building up, would justify ever increasing budgets for equipment and manpower, in a self-perpetuating cycle analogous to the "arms race" that also characterized that post-war era. It was that political and institutional environment that Harold Biswell contended with in the late 1950s.

NOTES

1. Stewart H. Holbrook, *Burning an Empire: The Story of American Forest Fires* (New York: Macmillan Co., 1943), Foreword by William B. Greeley, vii, viii.
2. Donald Culross Peattie, *A Natural History of Western Trees* (New York: Bonanza Books, 1953), 178–179.
3. Sikorsky R–4 and R–6 helicopters were used by the Army Air Corps in the Philippines and China/Burma theaters. After World War II they were immediately adapted for use in fighting forest fires in southern California. See Robert W. Cermak, "Fire Control in the National Forests of California, 1878–1920" (M.A. thesis, California State University, Chico, 1986), 532–534.
4. John McDowell, "The Year They Firebombed the West," *American Forests* (May/June 1993): 55.
5. H. N. Wheeler, "Controlled Burning in Southern Pine," *Journal of Forestry*, 42 (June 1944): 449.
6. Agnes C. Laut, "The Fire Protection of the U.S. Forest Service," *American Forestry* 19 (November 1913): 711.
7. R. M. Conarro, "Fighting Tomorrow's Fires Today," *American Forests* (April 1939): 214.
8. Ovid Buster, "For a Public Will," *American Forests* (April 1939), 159.
9. George Vitas, *Forest and Flame in the Bible*, A Program Aid of the Coopertive Forest Fire Prevention Campaign sponsored by the Advertising Council, State Foresters, and the U.S. Department of Agriculture, Forest Service—PA-93, reprinted December 1961, p. 3.
10. Ibid. 9, 10.
11. Steve Nelson and Jack Rollins, "Smokey the Bear" (New York, Hill & Range Songs, Inc., 1952).
12. Ellen Earnhardt Morrison, *Guardian of the Forest, A History of the Smokey Bear Program* (New York: Vantage Press, 1976), 8.
13. See Harold K. Steen, *The U.S. Forest Service: A History* (Seattle: University of Washington Press, 1976), 281.
14. Nelson and Rollins, "Smokey the Bear."

15. Jay Heinrichs, "The Ursine Gladhander," *Journal of Forestry* (October 1982): 643.

16. Ellen Earnhardt Morrison, *The Smokey Bear Story* (Alexandria, VA: Morielle Press, 1995), 34.

17. Quoted in William Clifford Lawter, Jr., *Smokey Bear 20252, A Biography* (Alexandria, VA: Lindsay Smith Publishers, 1994), 369.

18. Dean Wohlgemuth, "Do Forest Managers Have a Smokey Bear Complex?" *Georgia Game & Fish* (February 1972): 6.

19. Lawter, Jr., *Smokey Bear 20252*, 371.

20. Richard J. Vogl, "Smokey's Mid-Career Crisis," *Saturday Review of the Sciences* 1, no. 2 (March 1973): 23.

21. Robert W. Mutch, "Understanding Fire as Process and Tool," adapted from "Fire Management Today: Tradition and Change in the Forest Service," presented at Society of American Foresters National Convention, Washington, D.C., September 28 to October 2, 1975.

22. Harold Biswell, "Fire Ecology in Ponderosa Pine-grassland" in *Proceedings, Annual Tall Timbers Fire Ecology Conference, June 8–9, 1971* (Tallahassee, FL: Tall Timbers Research Station, 1972): 89, 94.

5. Harry the Torch

> "TORCHBEARER:" 1. a person who carries a torch 2. a) a person who brings enlightenment, truth b) an inspirational leader, as in some movement.

Whether or not Smokey Bear contributed to the atmosphere of controversy, those who endured criticism for their efforts to reintroduce fire to ecosystems developed considerable rapport, sometimes expressed as black humor. Harold Biswell received a short note written on May 5, 1958, by Raymond B. Cowles, biology professor at UCLA. Cowles had also stirred up foresters—another outsider daring to comment on "their" turf—by authoring, "Starving the Condor" in *California Fish and Game*.[1] He postulated that increasingly dense brush fields due to fire suppression might have contributed to the decline in California condor numbers. Cowles thanked Biswell for some reprints and added: "I am sure some of my conservationist friends will damn me (you are already damned!) when they read this plea for research." Biswell so enjoyed Cowles' characterization that he sent a copy of the note along to Harold Weaver.[2]

Later Cowles told Biswell that "Starving the Condor" had been "returned by the editors of all popular outlets with the statement that they did not dare publish it, even though by chance I might be right, and even the . . . editor of 'California Fish and Game' introduced my article . . . saying that these were my views and not necessarily those of Fish and Game."[3]

In a letter to Weaver in March, Biswell told him, "Our prescribe burning studies are moving along in good shape. More and more I become convinced that we will have to give far greater consideration to it, particularly here in California where the danger from wildfires is so great."[4] He was beginning to gather a cadre of supporters, including (not surprisingly) those ranchers, cattlemen, and sheep grazers who had been early practitioners of burning in California. Al Spencer, a "breeder and developer of the Romeldale Sheep," wrote Harold in November, after Biswell sent him a copy of one of his published articles, thanking him for enlightening naturalists and foresters, and "for the benefit of the World."[5]

More supportive correspondence came in April 1958 from Richard H. Pough of New York, president of the Natural Area Council, past-president of the Nature Conservancy and author of Audubon bird guides. Biswell revealed uncharacteristic pessimism about the long-term prospects for prescribed burning when he wrote back to Pough, but also showed that he understood the roots of the bureaucratic resistance that was proving so difficult to overcome: "Someone asked recently if I thought prescribed burning would ever be used more widely to reduce wildfire hazard in the way that we are using it. It might not be because some people think it would be difficult to administer. The present system is easy—no one wants wildfires and we do our best to put them out. They start from lightning or someone's carelessness during an exceptionally dry and windy period. If from lightning under these conditions, the fires are considered an 'Act of God' and are forgotten. This makes for easy management!"[6] He closed the letter by encouraging Pough to read not only the reprints and other material he was sending, but also Cowles' "Starving the Condors" article and Harold Weaver's articles on wildfires and prescribed burning in ponderosa pine.

Researchers studying the fire ecology of Australia, where early foresters had also instituted fire exclusion policies, began corresponding with Biswell and attended some of his field day demonstrations. Dr. J. E. Coaldrake wrote Biswell from Queensboro, Australia, in December 1959 saying "It's just a year since I spent that very interesting day with you at Hoberg's and my mind often goes back to your simple but very striking demonstration of the uses of prescribed burning in Ponderosa pine forest. It will interest you to know that I have met with troubles similar to your own in trying to start a program of research on fire. I have even been invited to find another job if I persist with this interest. The battle is not over and I have not given up hope yet." The remainder of the letter was a request for Dr. Biswell's comments on several research questions to help Coaldrake "keep gathering fresh fruits with which to back my case."

Biswell's "Reduction of Wildfire Hazard" article appeared in *California Agriculture* in May 1959. That month he wrote Weaver saying that soon he would be taking the state forester, the director of Natural Resources, the regional forester, and representatives from the forest industry on a tour of his Hoberg's Resort burn sites. "I don't know their purpose!" he exclaimed.[7] He also told Weaver about his upcoming article in the *Sierra Club Bulletin*, "Man and Fire in Ponderosa Pine in the Sierra Nevada of California." Biswell was beginning to reach out to a broader audience via the popular media.

Bruce Kilgore, who in another decade would lead the program to reintroduce fire within Sequoia National Park, had edited Biswell's *Sierra Club Bulletin* piece and was now editor for *National Parks Magazine*. He invited Biswell to write another article, this time on the relationship between fire and giant sequoias. "Such an article would, of course, also deal with the effect of current national park policy on both Big Trees and associated species." Kilgore mentioned that he had been a member of Dr. A. Starker Leopold's game management class that visited Biswell's experimental area in Lake County.[8]

Biswell very much wanted to write on that subject, but asked for time to complete some research on the Big Trees. "Concerning my article in the *Sierra Club Bulletin*, I think you will be interested to know that not a single adverse comment came to me about it. On the other hand, I received a good many favorable comments. I remember your field trip with Dr. Leopold to Hoberg's in 1952. I wish you could see the area now. I think you would be quite impressed."[9]

Clearly he was choosing not to hide his work away from the general public, nevertheless he did not want to alarm his more cautious superiors. So he asked the new Forestry School dean, Henry Vaux, to look over the *National Parks Magazine* article before publication. Vaux told him it was "a highly important subject and one with which all of us will undoubtedly be more and more concerned." It must have been gratifying for Biswell to read that Vaux thought "a good many foresters in California are thinking more open-mindedly about the role of fire in forest management than was true a decade ago. At the same time there are people who are less open-minded or who have some axe to grind in connection with fire problems. They will leave no stone unturned in an attempt to discredit material which might conceivably work against their particular point of view." The dean delivered familiar advice: that Biswell should be certain that "all material published by us on the subject of prescribed burning is couched in highly objective—even conservative—terms, so that no loose ends are left for critics to seize and unravel." He proposed informal review from other interested members of the department so that publications could be "unequivocally supported by the whole staff."

Biswell penciled a question mark in the margin below the words "unequivocally supported," perhaps questioning the stifling effect on groundbreaking work should universal preapproval be a prerequisite.[10]

In 1959 the California Division of Forestry published C. Raymond Clar's *California Government and Forestry*, a history of the state's forestry practices and issues "from Spanish Days until the creation of the Department of Natural Resources in 1927." Clar covered the light-burning controversy up through that early period and added a footnote recognizing how Dr. Biswell's research was readdressing the contentious issue. Biswell is "one of the most active advocates of the use of controlled fire at the present day," the footnote read. However, Clar was a graduate of the UC Berkeley School of Forestry who had worked his whole career with California State Forestry, and apparently counted himself as one of the professionals "whose reaction to such studies is reserved but courteous."[11] "It appeared to Clar that Biswell "makes no great distinction in fire used to enhance browse and fire used to eliminate a fire hazard under a high forest." Beyond that mild criticism, he dismissed Biswell's conclusions on the effects of fire upon the reproduction of various tree species as essentially no different than findings developed by others forty-five years earlier.[12]

Dean Vaux received a letter on March 23, 1960, from Knox Marshall, secretary of the California Forest Practice Committee (on Western Pine Association letterhead), expressing concern about pressures the logging industry was feeling for controlled burning of timberlands in California. The letter was addressed to the university dean because Dr. Biswell's work for the School of Forestry had begun to convince too many people, in the committee's opinion, that prescribed burning had proven its feasibility for reducing fuel hazards. The group sent representatives to the field demonstrations Biswell hosted at Hoberg's Resort in the spring of 1959, but remained unconvinced that controlled burning was applicable to large-scale commercial timber operations.

Though Marshall appreciated Dr. Biswell's work in meeting limited objectives at Hoberg's, he expressed doubt about costs and strong concern about risks that might follow any encouragement of controlled burning. His Forest Practice Committee had considered the issue in the past, Marshall wrote, but always concluded that disadvantages outweighed any possible benefits. The letter expressed the committee's support for more experiments, but cautioned against making overly broad generalizations in published results. Marshall closed with a final warning about high costs to timber productivity should well-intentioned promotion of fire as a tool encourage incendiaries among the general public.[13]

That autumn, despite such criticism, Biswell's "Danger of Wildfires Reduced by Prescribed Burning in Ponderosa Pine" appeared in *California Agriculture.*

Further complications in Biswell's relationship with the School of Forestry dean had been generated that summer by newspaper coverage of Dr. Biswell's views. The *Sacramento Union* published an editorial on August 26, 1960, headlined "Preventing Forest Fires." The opinion piece appeared while major wildfires were burning near Foresthill in the Sierra Nevada foothills east of Sacramento.

> Mounting evidence is at hand that controlled burning of brush will prevent disastrous forest fires. It comes from woodsmen, conservationists, sportsmen, all of whom credit controlled burning with different values, but who all unite in agreeing to its multiple benefits.
>
> Dr. H. H. Biswell, head of the department of forestry at University of California, insists that controlled burning can be conducted without danger to young timber stands. He adds: "If natural fire breaks had been created by controlled burns in Tahoe National Forest, the Foresthill and Donner Lake fires could have been extinguished before they seared 70,000 acres of state and national forest land and destroyed millions of board feet of timber."

Dean Vaux wrote the *Union* himself the next day to correct the misinformation about Biswell being head of the Forestry Department of the university and the quotation attributed to him "which Dr. Biswell tells me he did not make." The dean explained that more research was needed before prescribed burning could be used as a reliable tool for hazard reduction.[14]

Biswell followed up with a letter to the dean, agreeing with his call for more research and field-testing. However, by 1960 he had been conducting burn tests in ponderosa pine for nine years and was ready to push for practical applications. So he added:

> At the same time, I think we should make more use of the data we already have. Conditions suitable for both broadcast burning and cleanup burning have been well defined. Of course, there is always danger that fire will escape and do damage, particularly in those areas where the fire hazard has been permitted to reach high proportions. On the other hand, we have set fire perhaps a hundred times and have not had a serious escape yet.
>
> I feel we have made much progress and have gained much information that can be used and cited in matters dealing with the problem of fire hazard reduction in forest management. I want to suggest once again that members of the staff be urged to see the Teaford Forest as opportunity permits. I realize that they are very busy with their own programs, but still *I am amazed that not a* single *member of the forestry staff has seen the research on this forest* (at least to my knowledge) since the work was started there in the spring of 1951. This in view of the fact that some of them have been invited three or four times [italics added].[15]

Dean Vaux, though one of the "old guard," was open to cautious change. Aiming for a tone of constructive criticism, he replied to Biswell at length. Vaux speculated about why more use of Biswell results had not been made, taking the issue squarely back to California's light burning controversy at the beginning of the century:

> A good many of the people most directly concerned with forest management and wildfire prevention policies—the ones in a position to *use* your results—have not "accepted" your findings. . . . In my judgment there are several reasons. . . . One is that controlled burning is a very old concept that was tested and rejected almost fifty years ago. Even though your idea of controlled burning differs from the old concept, many people haven't yet learned the difference, and until they do they are unlikely to accept the new idea. . . . there is need to cultivate among forest managers a feeling of confidence in *you*. Your vigorous public advocacy of a policy which these men feel is wholly unevaluated from the economic standpoint is a major barrier to any attempt to cultivate the needed attitude of confidence.

Vaux mentioned, as an example, the letter from Knox Marshall, back in March, with the California Forest Practice Committee's concerns. The dean praised Biswell's work in calling attention to controlled burning. "But I also believe that continued progress is going to depend on securing the acceptance and cooperation of the men who have had lifelong experience in forest management and who bear very heavy personal responsibilities for it. I'm sure that such men will only be antagonized by broad generalizations that appear to ignore both the long years of careful study which many foresters have given to some of these problems, and the unanswered questions. . . . To secure this cooperation is going to require a very different approach than the one followed to date."[16]

Despite controversy within his own university school, the newspaper editorial was one signal that enough public support for Dr. Biswell's fire hazard reduction research was building to encourage him to extend his work into more of the Sierra Nevada forest types. A severe 1960 summer wildfire season set the stage for Biswell's proposal to University Dean Aldrich on September 30: "Suddenly, in recent weeks there has been a great increase in interest concerning this research, all generated by the tremendous wildfires of the past summer. With this interest it is my hope that funds can be obtained to *greatly increase* this research and demonstration." He was thinking in terms of $60,000 to $100,000 for research. "This seems like a large request but it is nothing more than 'a drop in the bucket' compared to the great expenditures and losses connected with wildfires."[17]

The proposal went to the dean of the College of Agriculture because it aimed to foster integrated research among various departments and interests in wildland resource management within the university. As was his inveterate habit, Biswell offered Dean Aldrich the opportunity to visit Hoberg's Resort and Teaford Forest to see the work on the ground. However, his prescribed burning studies would not be authorized in giant sequoia and mixed conifer forests until 1964.

In "The Big Trees and Fire" article for *National Parks Magazine*, Biswell used quotations that he considered important enough to incorporate into most of his speeches. "Very few statements," Biswell wrote, "can be found describing the actual behavior of fires in early times." However, John Muir had written a detailed description of a fire entering a forest of Big Trees from a brush field, between the middle and east forks of the Kaweah River. It had been the driest season of the year, when a fire occurred in early September:

> The fire came racing up the steep chaparral-covered slopes of the East Fork canyon with passionate enthusiasm in a broad cataract of flames. . . . But as soon as the deep forest was reached the ungovernable flood became calm like a torrent entering a lake, creeping and spreading beneath the trees . . . There was no danger of being chased and

> hemmed in, for in the main forest belt of the Sierra, even when swift winds are blowing, fires seldom or never sweep over the trees in broad all-embracing sheets as they do in the dense Rocky Mountain woods and in those of the Cascade Mountains of Oregon and Washington. Here they creep from tree to tree with tranquil deliberation, allowing close observation.[18]

Biswell went on to describe the giant sequoias' "asbestos-like bark," fire history derived from healed-over scars that indicated fires burned trees about every eight years without killing them, and lightning data and suppression records that showed "some fires must have been widespread in the Sierra Nevada *every year*." He quoted Galen Clark, first guardian of Yosemite Valley on the Indians' "management for their own protection and self-interest. . . . When the Valley was first visited by whites in 1851, [Indians would] annually start fires in the dry season of the year and let them spread over the whole Valley to kill young trees just sprouted and keep the forest groves open and clear of underbrush, so as to have no obscure thickets for hiding place, or an ambush for any invading hostile foes, and to have clear grounds for hunting and gathering acorns."

He also quoted Joaquin Miller, who wrote in 1887, "In the spring . . . the old squaws began to look about for the little dry spots of headland or sunny valley and as fast as dry spots appeared they would be burned. In this way the fire was always under control. In this way the fire was always the servant, never the master. . . . By this means the Indians always kept their forests open, pure and fruitful and conflagrations were unknown."

As always, while describing the consequences of decades of fire suppression, Biswell pointed to the essential need for wildfire control. But he also deplored the threat of uncontrollable holocausts now that "fuel ladders" of shade-tolerant firs and cedars had filled in the forest floor beneath the Big Trees. With continuous fuel from the ground to the canopy, crown fires would burn where they had once been almost nonexistent. Biswell had not yet begun his own research burning beneath Big Trees, but did describe the successful burns he and Harold Weaver had been conducting for years in ponderosa pine forests. Concluding, Biswell asked: "Would it not be worthwhile to select at least two groves of Big Trees—a dozen would be better—with their surrounding forests for some distance back, and manage them with light fires as a part of the environment, somewhat as they were managed by nature and the aborigines through thousands of years in the past? Soon it will become evident that a forest fire can be a 'friend,' as it was in aboriginal times, and not our worst 'enemy' as it is today."[19]

Biswell wrote the editor of *National Parks Magazine*, Paul Tilden, on June 5, 1961, about that article: "Every single comment has been favorable. For example, Henry Clepper, Secretary, Society of American Foresters wrote, 'Congratulations on your interesting article.... It is informative and very readable. More power to your pen!' Professor Eugene Lee, of the Political Science Dept. here told me... the idea of fire being a part of nature was something entirely new and fascinating to [him]."[20]

Tilden had anticipated critical mail, but told Biswell that the magazine had received very little; most reactions were "quite favorable." "The National Park Service people, however," Tilden added, "were quite unenthusiastic about the article... but we expected that."[21]

One positive reaction came in a letter from Arthur W. Hartman, the Forest Service fire control officer in the South who had reviewed his Georgia/California article three years earlier:

> I studied closely the thinking behind your treatment of Big Trees and Fire. While the details are different, in principle, the same changes appear to be occurring among the Big Trees of California as among the Little Pines of the rural south. Fuels are able to build up to quantities away beyond those of any known former time. If foresters hang back from supplying the required leadership towards exposing these developing conditions and coming up with corrective actions, the future costs in resources and reputations will be immense.
>
> As I add up your thinking, I must congratulate you as being a true Conservative.... That word has been used loosely, distorted and driven in reverse until its true applications and relationships have been about lost sight of. In the field of natural resources, Conservative and Conservation are Siamese Twins.
>
> A person cannot qualify as a conservationist and at the same time play free and liberal with the Laws of Nature... or take liberties with her Rules of Order. If you, I and some others are incorrectly reading and interpreting the ways nature has employed fire to develop and conserve forest resources, we are of course too dangerous to be left at large. If those interpretations are somewhat on target, individual foresters are going to have to choose between taking either positive or negative positions. If you want to spot some of those whose contributions will really be negative, look about for subjects who think they can outsmart nature.[22]

Albert A. Clapp sent Biswell a copy of a letter he had written to the Oakhurst Chamber of Commerce. Clapp's father homesteaded the foothills there, west of Yosemite, in 1889. He had provided his Chamber of Commerce with a copy of Biswell's Big Trees article, declaring that it was "a *Must* for Californians especially" and wrote:

> The first we knew of government forest protection in the North Fork area, was in 1897 when a fire was set along the Crane Valley (now Bass Lake) road to burn off a brushy side hill; a fire warden, the first we had seen, called for help to put it out. He got the help, but they nursed the fire along to the top of the ridge where they had intended stopping it; and some got paid for it. I was on that fire line.
>
> In 1910, after more than 12 years of keeping fire out of the forests, we old timers still held to the theory that cutting the low hanging limbs, clearing and burning trash, dead brush and downed timber any time when the grass is green is the only practical way to prevent the crown fires that take the green standing timber. While it has taken some people 50 years to learn these facts, we are glad to know that many able scientists and others, after witnessing the disastrous and practically uncontrolled conflagrations . . . are advocating this system, and while we have made good progress, we need help to reach the powers that be.[23]

Finally, Biswell was carrying on the long-term, systematic research that had never been done during California's early light-burning debates. With Harold Weaver's relevant work on ponderosa pine forests of the Southwestern states, there was mounting scientific support, at last, for ranchers and timbermen opposed to fire exclusion policies.

Yet the old guard was not prepared to capitulate. T. V. Arvola, deputy state forester in California and an alumnus of UC Berkeley, sent a letter to the *California Monthly* in May, 1961:

> Your News brief about forest fire control under University Panorama in the February, 1961, issue leaves a prejudiced impression with the uninitiated. Dr. Biswell's premise that light controlled burning would reduce wildfire hazards in the Sierra Nevada is not shared by his illustrious colleagues on the faculty of the University of California School of Forestry. Neither is it accepted by practicing professional foresters or more important, by timberland owners and managers who have the largest stakes in the matter.
>
> Light burning is impractical and uneconomic except in very special situations, often causes more damage than it would prevent and carries with it an onus of liability. In my opinion there has emanated too much

publicity from Dr. Biswell on this subject which is based on inadequate research and practice.[24]

The summer of 1961 found Biswell in Europe, funded by a Guggenheim grant and Fulbright award allowing him to study the role of fire in the Mediterranean regions of France, Spain, and Italy. The next year he went to Greece for further research and to teach range management classes in a Greek university. In his application for the Guggenheim grant, Biswell disclosed hopes to ultimately extend such research to all of the Mediterranean-type climate regions of the world, including Australia, North and South Africa, and Chile. Biswell gave Weaver's name as a reference in the grant application. By then the two had known each other for ten years.

Weaver explained that background in the comments he sent to support Biswell's grant application, adding, "Dr. Biswell has a very strong and positive personality and is not afraid to discuss facts as he understands them. This has, at times, made him unpopular with foresters who believe that it should be universally understood that fire is an unmitigated evil in the forest. I believe," he added, "that Dr. Biswell has already made noteworthy contributions towards knowledge of the ecological role of fire."[25]

Biswell had just returned from Greece when an invitation arrived from the Southern California Section of the Society of American Foresters. Their annual meeting was held December 1, 1962, in Oakland, and Biswell joined a panel there to discuss the use of fire as a tool in forest protection and management. Earlier that summer a wildfire had burned into Hoberg's Resort onto the lands he had been treating for many years. That incident served as a direct test of the fire protection afforded by prescribed burning. Foresters' skepticism and outright defiance revealed in the record of that meeting show how pressure for change, supported by the new research, was generating resistance within the forestry profession.

A policy statement adopted in April by that regional section of the Society of American Foresters was the focus of the panel discussion. That policy declared that

- present knowledge was inadequate to employ fire as a tool;
- fire should only be used when no equally promising alternative techniques could meet burning objectives;
- public assistance for burning on private lands was not warranted;

- use of fire would not be condoned unless prescribed by a professionally qualified person;
- generalized statements concerning the efficacy of fire as a solution to all Wildland problems everywhere were misleading, dangerous and should be challenged at every occasion.[26]

In his portion of the panel presentation, Biswell told the gathering that the wildfire near Hoberg's in late August that year was "nearly a perfect demonstration" that prescribed burning could greatly reduce the hazard and danger of wildfires.

> One of the headfires came directly into one of the areas that had been prescribed burned. On the outside it had been crowning and there it burned off the needles and killed every tree in its path. But when the fire reached the area that had been prescribed burned it went to the ground, became relatively calm, and progressed rather gently through the pine needles on the ground. The fire acted much like one in prescribed burning and it may have done more good than harm by removing debris and further reducing the danger of wildfires. At present [4 months later] this area is well covered with a new crop of pine needles and in places it is difficult to see that it ever burned, while outside the area the soil is nearly as bare as the floor of this room and surface runoff has been very heavy.[27]

He also detailed the research studies that had been completed and suggested that such work be expanded to include coast redwood, Douglas fir, Big Trees, and mixed conifer forests, "where severe fire hazards now exist. . . . Research and demonstration need not hold up an action program. I would recommend that an action program be started immediately by typing areas of ponderosa pine where prescribed burning can be done and by training personnel to do the work."[28]

The society's policy statement itself and the discussion by others on the panel showed how far Biswell's attitude was from the entrenched California foresters'. Assistant Deputy State Forester L. T. Burcham, with the CDF, led the attack, saying that there was "no reliable experimental evidence that these objectives—or the greater part of them—can generally be attained."[29] Devaluing the research of Dr. Biswell and Harold Weaver, Burcham maintained that it was a "fact that sound research information . . . to conduct burning safely, to predict results accurately, and to assess its economic values correctly, simply does not exist."[30]

One of Burcham's issues was that "professional foresters" (a phrase repeatedly emphasized in his remarks) had not conducted the necessary research, and recommendations from outside the profession were, clearly, neither welcome nor greatly valued. Dr. Biswell, though on the faculty of the School of Forestry at UC Berkeley, held degrees in ecology, rather than forestry. The only other member of this panel speaking favorably about prescribed burning was R. Merton Love, chairman of the Department of Agronomy at UC Davis, also not a "professional forester." Burcham stated:

> For more than sixty years *professional foresters* have provided leadership in the conservation and development of natural resources. . . . Collectively, they have accumulated an impressive total of man-years of actual experience in managing forests and other wild lands. This experience . . . applied with *professional* judgment and skill, cannot be surpassed or even matched by any profession or group concerned with conservation and administration of natural resources. Yet, interest in extending the use of fire in wildland management has reached such proportions among the general public that *professional foresters* are being criticized for not embarking upon a program of planned burning. . . . To a great degree this criticism is stimulated by individuals who have neither formal training nor practical experience in either the management of wild lands or the use and control of fire; who have no responsibility for management of wild lands; and who lack an understanding of the fundamental ecological principles involved. Current proposals for widespread use of planned burning . . . are based almost wholly upon superficial observation and personal opinion; they do not have a sound basis in scientific research, and are at variance with the general experience of *professional wild land managers* [italics added].[31]

Burcham closed his statement, as opponents of prescribed fire often had, with a call for more research. In the discussion that followed, answering questions from the audience, Burcham amplified his emphasis on outsiders who dared to criticize practices of the forestry profession: "To illustrate the point that I've tried to make with a contrasting statement: I think that we very well know the hell that would be raised if we foresters go out and start recommending to grain farmers that they burn their barley in order to get their wheat."[32]

Questions directed to Biswell focused on costs and economics, soil fertility, timber site qualities, and rodent responses following prescribed burns. Finally, his interpretation of the effects of the wildfire was questioned. Had he taken into account the time of day, terrain, fire behavior and weather as the wildfire approached his treated area?

Biswell answered that the August fire reached the plot about 6:00 PM. "This treated area was directly in the path of the fire. When it got to this treated area, then, the fire went down onto the ground, but over to the left side of this, and also to the right side of it, it killed every tree. . . . But the head of the fire went right square into the treated area. There it dropped to the ground, but where not treated, it crowned through trees. Some say the wind velocity went down [when it hit the treated area]. Maybe that's so, but then the other report is that the velocity over the entire area went down."[33]

The panel moderator asked Chuck Fairbank to comment from the audience at that point, as he had been on the wildfire scene: "Dr. Biswell's statement doesn't bother me a bit. I think the heat that turned into the prescribed area hit a bench just before it hit this area and came down. . . . Basically, I think as a forester and one who is interested in fire control, I think we have to be awfully careful in making a statement that this prescribed burning stopped the crown fire and put it on the ground. *This is not so*" [emphasis in original transcript].[34]

Years later, Dr. James Agee, a professor at the University of Washington who had been one of Biswell's graduate students at Berkeley, described this exchange in a memorial speech about Biswell: "Personnel from the fire suppression agency involved testified that the wind stopped exactly at the edge of the prescribed burn unit, so that a change in weather was responsible for the change in fire behavior. They were probably right that the wind slowed, but it slowed because the prescribed burned area had a dampening effect on the wildfire's behavior. I was able to visit the site years later, and I found all the trees dead in the wildfire area and a healthy forest in the prescribed fire area. An objective analysis was sorely lacking."[35]

What should have been obvious, given an objective analysis, was flatly denied, however. Professional foresters were feeling pressure for change and their resentment found a target in Dr. Biswell, in particular, whose speeches and articles in both academic and popular journals fueled that pressure for change. Harold Weaver commiserated with Biswell, after reading about the exchange. "A friend, who also read the material," Weaver told him,

> didn't like L. T. Burcham's presentation and suggested that I take issue with him. I have never met the man, though I recognize his type. It

would avail little to argue with him, though I do resent his insinuations that we have drawn conclusions solely from rather narrow observations and experiences and have just been lucky. Has he seen your prescribed burning tests at Hoberg's? Perhaps he is like the old farmer who, when he first saw an elephant at a circus, said, "There just ain't no such animal."

Your California foresters appear very touchy about any information being given the general public on . . . fire as a management or hazard reduction tool. Apparently the general public is too ignorant to be trusted with any reasonable discussion on this subject.

Aside from sour notes coming from Burcham, I am much encouraged by the papers that you sent me. Even Burcham calls for research on the subject, and I would call that "progress."[36]

Biswell's difficult, most controversial years of work were about to surrender, in fact, to a satisfying groundswell of support that, in the 1960s, would open the way to changing fire exclusion policies.

Both opponents and supporters had begun calling Biswell "Harry the Torch." The nickname, referring to the drip torch used to ignite fires while prescribed burning, was a condemnation in some circles, but was also used with respectful affection by many admirers. In a time of generational rebellion against the nation's Vietnam war policies, the aging university professor would become, in the late 1960s, a surprisingly charismatic "rebel leader" himself, through speeches, publications and demonstrations—peaceful ones—at his study sites. He was, in fact, a "torchbearer," fulfilling each of that word's definitions: "1. a person who carries a torch 2. a) a person who brings enlightenment, truth b) an inspirational leader, as in some movement."[37]

In the 1960s and 1970s much controversy would still have to be endured, but it would be shared among land management bureaucracies and their personnel, activists in this "anti-war" movement to "give peace a chance" in the war against nature's fires. "Harry the Torch" would extend his burning research into Sierra Nevada pine and Sequoia forests and, after Harold Weaver retired from the BIA in 1967, the two "Harolds" would collaborate on fire ecology studies of the giant sequoias. Their concern for the Big Trees would help bring fundamental changes in fire policy, first in parks protecting groves of the rare giant sequoias, but ultimately extending to other wildland ecosystems and other land management agencies. A major turn-around point arrived in 1962, in part because that year the Tall

Timbers Research Station initiated an influential series of national fire ecology conferences.

NOTES

HB—Harold Biswell papers, held at the Bancroft Library, University of California, Berkeley (BANC MSS 2002/67 c).

HW—Harold Weaver collection, held at the Forest History Society, Durham, North Carolina.

1. Raymond B. Cowles, "Starving the Condor," *California Fish and Game* 44 (1958): 175–181.
2. R. B. Cowles to Biswell, handwritten note, May 5, 1958, HB.
3. R. B. Cowles to Biswell, May 18, 1961, HB.
4. Biswell to Weaver, March 26, 1958, HB.
5. Al Spencer to Biswell, November 12, 1958, HB.
6. Biswell to Richard H. Pough, July 16, 1958, HB.
7. Biswell to Weaver, May 4, 1959, HW, Box 2, File 14.
8. Bruce Kilgore to Biswell, January 27, 1960, HB.
9. Biswell to Kilgore, February 2, 1960, HB.
10. Henry Vaux to Biswell, September 18, 1958, HB.
11. Clar was chief deputy state forester from 1941 until 1953, then assistant executive officer for the State Board of Forestry.
12. C. Raymond Clar, *California Government and Forestry: From Spanish days until the creation of the Department of Natural Resources in 1927* (Sacramento: California State Board of Forestry, 1959), 492–493.
13. Knox Marshall to Vaux, March 23, 1960, HB.
14. Vaux to *Sacramento Union*, August 27, 1960, HB.
15. Biswell to Vaux, memorandum, September 7, 1960, HB.
16. Vaux to Biswell, September 13, 1960, HB. Dr. James K. Agee, one of Biswell's graduate students in 1969, wrote me on July 11, 2001, that Dean Vaux was "a cautious social scientist" and "one of the 'old guard'" but he "did turn around"; he gave Agee personal support "on keeping up the pressure on fire issues."
17. Biswell to Aldrich, September 30, 1960, HB.
18. John Muir, *Our National Parks* (New York: Houghton Mifflin, Co., 1909), 307.
19. Harold Biswell, "The Big Trees and Fire," *National Parks Magazine*, April 11–14, 1961, 14.
20. Biswell to Paul M. Tilden, June 5, 1961, HB.
21. Tilden to Biswell, June 15, 1961, HB.
22. Arthur W. Hartman to Biswell, May 9, 1961, HB.
23. Albert A. Clapp to Biswell, May 30, 1961, HB.

24. A copy of page from the May 1961 *California Monthly* with Arvola's letter was sent to Biswell by Walter Emrick of the Madera County Agricultural Extension Service after Emrick received a letter from the California Wool Growers Association saying "We hope the enclosed can be answered by you. T. V. Arvola shouldn't get by with this." June 12, 1961, HB.

25. Undated, but also see Biswell to Weaver, October 16, 1959, and Weaver to Biswell, October 22, 1959, all in HW, Box 2, File 14.

26. Society of American Foresters, Southern California Section, "Policy Statement on the Use of Fire in the Management of Forests and Other Wild Lands in California," in *Proceedings, Society of American Foresters, Southern California Section, Annual Meeting, December 1, 1962, Oakland, California* (1962), 11.

27. Society of American Foresters, "Fire as a Tool in Forest Protection and Management," in *Proceedings, Society of American Foresters, Southern California Section, Annual Meeting, December 1, 1962, Oakland, California* (1962), 7.

28. Ibid., 9.

29. Ibid., 12.

30. Ibid., 13.

31. Ibid.

32. Ibid., 19.

33. Ibid., 23, 24.

34. Ibid., 24.

35. James K. Agee, "Memorial Dedication to Dr. Harold H. Biswell," in *The Biswell Symposium: Fire Issues and Solutions in Urban Interface and Wildland Ecosystems, February 15–17, 1994*, USDA Forest Service Gen. Tech. Rep PSW–GTR–158, Walnut Creek, California, Pacific SouthWest Research Station, Albany, CA, 1995, p. 1.

36. Weaver to Biswell, February 20, 1963, HW, Box 2, Folder I.

37. Definition from *Webster's New World Dictionary*, 2nd College ed. (New York: Simon and Schuster: 1980).

PART II: WHO WERE THE ANTI-WAR ACTIVISTS OF THE 1960s AND 1970s?

The practical man is the adventurer; the investigator, the believer in research, the asker of questions, the man who refuses to believe that perfection has been attained. . . . There is no thrill or joy in merely doing that which anyone can do . . . It is always safe to assume, not that the old way is wrong, but that there may be a better way.

Henry R. Harrower (a long-hand note in Dr. Harold Biswell's papers)

6. Tall Timbers

> Few people have ever stirred such an awakening in the ecological sciences; fewer still have pressed onward to see their efforts applied to wildlands across so many regions of the earth.
>
> J. Larry Landers, describing E. V. Komarek, 1991[1]

The words "fire" and "ecology" were first joined together in 1962 for the first national Fire Ecology Conference in a series sponsored by the Tall Timbers Research Station in Tallahassee, Florida. Ed Komarek, the conference founder, described the opening conference as "the very first professional meeting in the United States to fully address the beneficial effects of appropriate burning upon natural resources."[2] Conferences brought together individuals active in conservation and natural resource management to better understand the use of fire and its role in nature and to stimulate fire research worldwide.

Ed Komarek had joined Herbert Stoddard's Cooperative Quail organization as his assistant on July 1, 1934. After the quail association disbanded in 1943, Stoddard and Komarek did similar work privately, using fire for wildlife, vegetation, and agricultural objectives. In 1939 Ed and his brother, Roy, acquired property adjoining Stoddard's Sherwood plantation and near the Tall Timbers plantation. Only 75 of 565 acres of the property were tillable at the time, but they reclaimed the land with controlled burns. In 1945 Ed and Roy also took on management of the Greenwood Plantation. Tall Timbers Research Station was founded in 1958.

Ed Komarek at the Tall Timbers Research Station, Tallahassee, Florida. Courtesy of Tall Timbers Research Station.

Komarek was born in Chicago on June 4, 1909. "In his youth Ed was strongly influenced by the fire exclusion movement and, as a boy scout, volunteered to help beat out wildfires in nearby preserves."[3] He joined the Chicago Academy of Science after high school as a museum mammologist and studied zoology at the University of Chicago. He was surveying mammals through the South when he met Herbert L. Stoddard, Sr. "His unbounded enthusiasm became focused with the founding of Tall Timber Research Station, an institution conceived years before by Stoddard, and funded by friend and benefactor Henry L. Beadel, who had been heavily involved in the Quail Association."[4]

Harold Biswell submitted a paper for the 1962 fire ecology conference, where he summarized his years of wildland fire ecology research in California. In his presentation Biswell gave credit to Stoddard's articles as some of the first information available to help guide the burns Biswell conducted in Georgia. "I can say," Biswell added, "that [Stoddard] is a Master in the *Art* of controlled burning in the pine forest of the deep Southeast."[5]

Komarek wrote to Harold Weaver in April 1962 because he was touring the West to prepare for the second Tall Timbers Fire Ecology conference (to be held the following spring). The Tall Timbers staff had followed Weaver's burning in the western states. Komarek wished to meet both with Weaver and Harry Kallander.

In December Komarek invited Weaver to speak at the 1963 Fire Ecology Conference. "I have been much impressed, both by your writings, as well as the practical application of fire on Apache, Klamath and Colville Reservations. The many objections raised by some foresters to the use of fire in Ponderosa seem awfully reminiscent, even to the actual phrases used, to what Mr. Stoddard and I have had to put up with, until recently. Now one would think the Forest Service invented the use of fire in Longleaf and Loblolly."[6] Komarek mentioned he had hoped to see Dr. Biswell during his western tour, but found out when he got to Berkeley that he was in Greece.

Weaver requested BIA permission to attend the conference. His letter to his supervisor revealed concern that the request would be denied; Weaver made it a personal plea: "I have firmly in mind your letter to me of August 31, 1962, and my reply of October 5. In it I called your attention to the fact that my past advocacy of more research on fire in ponderosa pine has frequently made me the center of attention with respect to this subject." He was aware of the "austerity program" within the BIA at the time, but "I would be anything but frank if I did not indicate that I would very much like to go to the conference. I have always wanted to visit the Southern Pine region and this looks like the best chance, if not the only chance, that I may ever get, for I am nearing the end of my career as a forester. I will not harm the Bureau's relationship with anything that I may present. In fact, I may do them some good."[7]

Weaver, on the next day, also sent Komarek word that Biswell was back from Greece. Komarek, assuming that Weaver would attend, told him, "I am now hoping that we can get Dr. Biswell to also discuss fire and [southern] California at the conference. I heard some vague references among forest service personnel that he had had some unhappy experiences because of his views, probably similar to what Mr. Stoddard and I had to contend [with] in years gone by."[8]

Weaver was denied permission to attend. He sent that disappointing news to Komarek on January 30, 1963: "The official letter refusing permission showed quite plainly that the Office does not want me to discuss the ecology of fire in ponderosa pine under any circumstances. After enumerating the various jobs that I am expected to do this spring, the letter suggests quite pointedly that there will not be time available for me to go to Tallahassee."[9]

Komarek had to wait a week to let his anger cool before he wrote back. "We are mighty concerned about you not being allowed to attend, even at our expense and your time. Frankly, if you were not so near retirement I would force the issue" by seeking support from friends who were senators. "I had been somewhat afraid of this," Komarek continued. "Mr. Stoddard and I had hoped that this sort of thing was behind us. You see some 25 years ago I was even threatened with arrest for burning a client's land with his express permission. I would have written you earlier but I am still a bit hot under the collar. Mr. Stoddard and I have leaned over backward to be more than fair with the various services. However if some of them continue to try un-American tactics we can sure have a good discussion in Congress when appropriation bills come up for hearings."[10]

Harold Weaver *was* allowed to attend the Third Tall Timbers Fire Ecology conference in 1964, where he spoke on "Fire and Management Problems in Ponderosa Pine." He had twenty-one years of experience burning forest lands by then. Illustrating the magnitude of the problems caused by six decades of fire exclusion, Weaver told the gathering that in the Northern Region of the Forest Service, 47 percent of the total commercial forest area (7.6 million acres) had been taken over by dense pole and sapling stands.[11]

After the conference Weaver wrote Biswell, enthusing about the experience. It was "for me, the realization of many years of dreaming of someday visiting the Southern Pine Region and of visiting forests where fire is used intensively." His presentation paper, though, had been somewhat modified by his agency. "My original conclusion was that an action program is needed in ponderosa pine. It was changed to indicate that more intensive research should be pursued. . . . I can't quarrel with this, but I have been calling for research for many years. I personally think that we have enough to enable us to take some real corrective action, such as has been taken on the Fort Apache Reservation. The plea for more research can be used as an excuse for doing nothing." While in Florida, he visited biologist William Robertson at Everglades National Park, where 5,800 acres of slash pine had recently been burned in the first prescribed fire program for that agency. Weaver added, "I am much interested in your research project in the Big Trees. I would say that 'its about time.' "[12]

Harold Weaver inspecting heavy fuels at Whitaker's Forest, July 28, 1966. Photo by Harold Weaver, BIA forester.

Weaver was a talented photographer who never traveled without his cameras. He would process his own negatives in his darkroom, then use government facilities to prepare black-and-white enlargements that he turned into official photo-essay reports of his field trips. He also habitually made photographic gifts. Herbert Stoddard returned from a trip in late May and "found that beautiful picture of the virgin Longleaf that you took during our recent Fire Conference. Have had the picture framed and it will decorate our walls from now on. I have never seen such a beautiful picture of this pineland, and will treasure it as long as I live. The Komareks left about a week ago for Mexico," Stoddard added, "where Ed will be gathering data on fire use, you may be sure."[13]

In 1964 Harold Biswell began giant sequoia fire restoration studies at a 320-acre experimental forest owned by the University of California. Whitaker's Forest, just west of Kings Canyon National Park,

Harold Biswell inspecting built up fuels at Whitaker's Forest, July 28, 1966. Photo by Harold Weaver, BIA forester.

had more than 200 mature, old growth giant sequoias. The forest was part of the Redwood Mountain grove of Big Trees extending into the national park. "When work began in this area wildfire hazards were extremely high," Biswell told California agronomists nine years later. "It had been judged unwise to take 4-H Club boys and girls camping there because of the danger of wildfire. The forest floor was littered with about 44,000 pounds of debris per acre." Beneath the larger trees, every acre held 500 dead and standing small trees and 900 living white fir and incense cedars, one to eleven feet tall. The forest was nearly impassable in places. Most browse and food plants for wildlife—and most wild animals—were gone from the area. "Through careful and sensitive manipulation and burning, portions of Whitaker's Forest are taking on open and park-like qualities and shrubs and wildflowers are returning. The forest is becoming scenic. As of now it appears that burning will not have to be limited to 'cool' conditions but that it can be continued throughout the dry summer period. After all, this is when most of the lightning fires occur and when the forest burned naturally before white man interfered."[14]

In the mid-1960s Biswell introduced fire ecology courses into the UC Berkeley curriculum, emphasizing the role of natural fire and use of low intensity fires in vegetation management. Before that time forestry courses focused on fire control and protection against fire. During the summer of 1964 Biswell sent Weaver an upbeat letter about the attitude change that seemed to be spreading: "The Forest Service here is getting more interested in using fire. . . . In fact I think everyone around here is getting a little more interested in learning about fire. What about getting Komarek (Tall Timbers) to finance you on a trip to get more pictures in Ponderosa pine [as] he wants a good book on the subject? He might be glad to do this. Perhaps you should stick by your job for another couple of years to see how things work out. Maybe things will get better! At least, there seems to be a greater interest in fire."[15] In that letter, Biswell also mentioned, "Hartesveldt is planning to burn some 2-acre plots in the Nat. Parks in August."

Any prescribed burning in a western national park was *big* news. Emanuel Fritz, in applauding Biswell's work at Whitakers giant sequoia forest, complained that he had called on the National Park Service "for more than twenty years" to do something about "the situation that is becoming worse as the competing trees grow larger and more demanding of soil moisture. But its policy of 'letting nature take its course' inhibited interest and action."[16]

Yet new action was cautiously beginning at Sequoia National Park, where Richard J. Hartesveldt of San Jose State University had been conducting a five-year research study funded by the National Park Service. Burning actually did not happen that summer, but in a December 1964 article for *Natural History* magazine, Hartesveldt described what he (and Thomas Harvey) planned for small study plots. "There is little doubt that careful use of fire and cutting constitute a much more realistic approach than does a policy of 'hands off.' "[17]

Forest Service lands were also receiving some small scale prescribed burn experiments in California. Jim Murphy and Harry Schimke of the Forest Service's Pacific Southwest Forest and Range Experiment Station had begun projects on the Stanislaus National Forest. And Dr. William R. Beaufait, from the Fire Research Laboratory in Missoula, Montana, was a "leading exponent of prescribed burning in the west" according to Weaver. He told Komarek that Beaufait "suggested to me that, regardless of whether or not they know it, or admit it, the Californians' interest has been stimulated by the work of Harold Biswell."[18]

A sense of near euphoria permeates the correspondence of these men during that period. "I note your comments on Biswell," Komarek wrote to Weaver. He further noted:

> The use as well as the ecology of fire has suddenly become nearly the "fashion" from one end of the country to the other—and I really shouldn't limit it to that for our correspondence is world-wide. This past year has seen the switch of emphasis and interest in fire and fire ecology from just the forestry agencies—now anthropologists, archeologists, geographers, etc. are all asking for information. . . . It is going to be very difficult for forest agencies to mislead the academic field much longer—and that may make some of the federal agencies less autocratic. And by the way we have had requests for our publications and some nice comments from California District Rangers!!!!!![19]

THE ODD COUPLE

Weaver traveled to California the summer of 1965 and, in his honor Biswell organized a week-long field tour that included Hoberg's, Whitaker's Forest, Yosemite, Stanislaus National Forest, and other sites. Weaver later produced one of his photo-essays to document that tour.

January 1966 brought Biswell a request from the superintendent of Yosemite National Park to present a slide presentation on control burning to the park staff. In March of that year Biswell, Weaver, and Harry Kallender were on the program of the Department of Interior's Fire Training School in Denver to speak about hazard reduction by fire. Weaver attended and spoke at Biswell's spring field day at Teaford Forest that May. Afterward Weaver went to Whitaker's with Biswell. In a letter to Paul Keen, Weaver said, "The understory has been cleaned out and disposed of by burning on two 20-acre tracts. . . . It looks good, but there still remains the problem of reducing the heavy accumulation of duff. Conditions were ideal for burning of this material, but Harold and his co-workers were afraid because the honor-camp inmates left the woods at 3:30 each afternoon, just when they would be most needed in conducting such critical work. It's too bad he can't get assistance from the Park Service with a real fire crew and adequate equipment."[20] Weaver, planning to retire early in 1967, was in California a lot that final summer of his BIA career. He again participated in a Biswell field day in July, this time at Whitaker's Forest.

Commenting on a report prepared by Weaver on the potential for intensive timber management on Indian reservations, Biswell wrote him a letter that seems a bit formal, considering how much time the men had been spending together. It may have been written primarily for the eyes of Weaver's superiors, to help Weaver convince the BIA of the need for action instead of just more research. Biswell must have known very well that Weaver agreed with everything he wrote:

> I would like to make a few comments and suggestions . . . and I hope you will not be offended. With your own experiences, Kallander's and mine in prescribed burning in Ponderosa pine, I am convinced that what we need most is an action program in using light ground fires to reduce hazards and to effect some thinning. We don't need more research in order to start an action program. Certainly, we know enough from our past research to go ahead and recommend prescribed burning as a desirable tool to reduce ground fuels and to thin out suppressed slow growing trees in thickets. Probably your greatest need at this time is not more research, but the selection of men with patience, intelligence, and alertness, (and good pay) to serve as leaders of small crews to go ahead with this work. Best regards and sincerely, Harold Biswell.[21]

Harold Weaver retired in June 1967, ending a thirty-nine-year career with the BIA. He was presented the Department of Interior's highest honor, the Distinguished Service Award, in recognition of an outstanding career as a forester with BIA, signed by Secretary of the Interior Stewart Udall. It read (in part): "Mr. Weaver was a pioneer in assessing the use of prescribed fire as a management tool in ponderosa pine forests. Today he is a national leader in this field, and through his efforts the Bureau has gained national and international recognition. Mr. Weaver's writings are valuable references, and are used by many universities in teaching forest management courses."

Weaver's comments in a letter to Ashley Schiff (after the 1962 publication of *Fire and Water: Scientific Heresy in the Forest Service*) reveal his thoughts a few years before he actually retired about the career and controversy he had faced:

> From my point of view, there certainly has been close adherence to the gospel as it is enunciated by the Old Guard of the Forest Service here in the West. Many individuals of the Forest Service, including Forest Research, have agreed privately with my point of view regarding the possible use of fire in ponderosa pine silviculture and protection. When it comes time to speak up in forestry and advisory meetings, however, they simply say nothing. They haven't dared to.
>
> For many years I was an admirer of the crusading zeal of the Forest Service and of its apparent esprit de corps. I still am to a certain extent. It took me a long time to realize that there are some very adverse influences on the reverse of the coin. These definitely stifle initiative and assure conformity. Perhaps this is inevitable in a large bureaucratic organization, but I had hoped that there might be exceptions to this rule. I am glad now that, through pure chance, I never accepted a permanent appointment with them. I would never have been able to do what I have. Also, for my heretical utterances, I probably would have been fired long ago, or banished to Siberia.[22]

Retirement released Weaver from constraints he had felt throughout his agency career. He explained how that felt to Ed Komarek the day before he left on a ten-day consulting job for Biswell and the University of California:

> Don't know exactly what Harold has in mind, but imagine that it will involve photography and the work will be in the vicinity of Whitaker's Forest and Grant Grove. In the light of your suggestions I have been giving my proposed paper [for the next fire ecology conference] considerable additional thought. It is true that I always have had the Bureau of Indian Affairs looking over my shoulder, and have had to specialize in the "soft sell." That situation no longer obtains, and I shall have to get used to it. I can expand my paper and say the things that need saying. . . . I shall work on it following my return from California, and before the summer has passed shall send you a revision.[23]

One of Harold Biswell's former students, Jim Agee, remembered Weaver's trips to California to work with Biswell. "Weaver was quite eccentric; very quiet, and quite different from Harold Biswell who was so ebullient and outgoing. The two of them were really kind of 'the odd couple' together—one person who was very active and running around and the other person was quite quiet and content to do writing and take his photographs. Weaver was never much of a public speaker. He was a very quiet man, but quite profound," Agee said.

"He would return from work in the forest and immediately take off his shoes and walk around barefoot. And then he would go and find a dark spot and work on his camera, taking out the negative and putting in new film. Both Harolds used 4-by-5 cameras that allowed for a very large negative and very sharp quality photograph."[24]

The close collaboration by the two Harolds and the freedom Weaver was given by retirement led to a co-authored article, "Redwood Mountain," for *American Forests* in August 1968. The team would also produce "How Fire Helps the Big Trees," published in *National Parks Magazine* in 1969. As much space was devoted to Weaver's photographs in these papers as to the narrative.

Jim Agee was one of a second generation of prescribed burn pioneers who were trained or influenced by "Doc" Biswell (as students called him) and then moved into responsible positions with agencies. Agee ultimately became a professor of forest fire ecology at the University of Washington. In 1966, working on his bachelor's degree at UC Berkeley, he took Biswell's range management course. Agee worked at Whitaker's Forest in the summer of 1967 monitoring air pollution emissions and doing some pile-burning. When Biswell began annual workshops at Whitaker's, he had Agee and another graduate student, Jan van Wagtendonk, handle various parts of the presentations. "He would have us take a ten or fifteen minute segment to give us a little practice talking about fire," Agee recalled. He remembered Biswell showing people that they could stick their hands directly beneath burning fuel, under the duff. "The National Parks advisory board came out to Sequoia and attended one of Doc's workshops out there, in 1967. At that time there was kind of a ground swell of support for the kinds of things he was talking about; [agency opponents] tended to fall into the background. By the end of the '60s, that was probably the watershed; I think Doc had most of his professional problems behind him. . . . There *were* still members of the CDF who would attend the workshops and kind of sit there stone faced."

Agee attended the first western Tall Timbers conference in 1967 organized by Biswell and Komarek at Hoberg's Resort. "That was an eye-opener. People from the South would jump up and declare, 'I believe in prescribed fire.' And somebody else would jump up . . . very much like a revival meeting." Agee found the southern style jarring after Biswell's more secular scientific style. The latter was the kind of science he aimed to practice in his career.

A QUEST FOR ECOLOGICAL UNDERSTANDING

Tall Timbers Research Station brought together a task force of experts to summarize ponderosa fire management in 1973. Biswell chaired the task force, which focused on the successes of the central Arizona Indian reservation burn programs. Their report, "Ponderosa Fire Management," was a culminating publication in the BIA careers of Harold Weaver and Harry Kallander, both recently retired. Two more task force members were Roy Komarek, from Tall Timbers, and Richard J. Vogl, biology professor at California State University, Los Angeles.

The Tall Timbers' task force study described BIA burning in Arizona on more than 93,000 acres of ponderosa pine forest since 1948. "The original Arizona ponderosa pine type was more of a savanna or parkland than a forest . . . with widely-spaced trees with a grassland understory or . . . groves of trees intermixed with pockets of grassland." Fire scars and ring counts showed a frequency of natural fires every six to seven years. "The maintenance of ponderosa pine . . . involved a mosaic of various-aged groups of pines, with each site or grove being cycled from meadow-opening to pines to meadows and back again by the driving force of fire. And under this natural management system, without any need for fire protection, thinning operations, or sanitation cutting, man found lumbering and grazing that exceeded anything comparable in his native Europe; quality resources that lumbermen and cattlemen would give their eye teeth to have again."

The booklet summarized the state of knowledge and practices and called for, as an urgent need, the development of public education programs like those used regularly in the South during the controlled burning season.[25]

The 1974 Tall Timbers Fire Ecology conference, the fifteenth in the series, would be dedicated to Harold Weaver and William G. Morris, to recognize their pioneering work in fire ecology and fire research in the Pacific Northwest. From 1970 through 1973 conferences were titled, "A Quest for Ecological Understanding," a favorite phrase of Ed Komarek's. Explaining a decision to end the fire ecology conference series after 1974, Ed Komarek said, "We feel we have accomplished our original goal, and that was to stimulate fire research on a worldwide basis. Fire research is now a common field of investigation by governmental agencies, universities, and many other institutions and individuals."[26] The conferences were too important and too well considered, however. Another, in 1979, was the last coordinated by Ed Komarek. The series began again, in 1991, held in alternating years from that point, with the twenty-first held in the year 2000. Some of the conferences were devoted to specific topics, including grasslands, wetlands, and high-intensity fire, or special regions, like Africa and Europe.

"More than anyone else, E. V. Komarek . . . promoted the concept of fire as one of nature's most potent evolutionary and ecological forces."[27] Komarek directed the Tall Timbers Research Station for twenty-one years. He ultimately delivered lectures in twenty-four states and fourteen countries and was awarded an honorary Doctor of Science degree from Florida State University. Komarek's papers were donated to Tall Timbers in 1987 and became the genesis of a fire ecology database, named for him, that holds over 12,000 records and can be searched on the Internet.[28] "I believe Ed Komarek could sell a forest fire to Smokey Bear," James Stevenson said at the 1989 Tall Timbers Fire Ecology Conference that honored Komarek.

By 1972 the University of California was no longer concerned about "keeping things quiet" about Harold Biswell's successes. They issued press releases instead, describing his demonstration burns at Whitaker's Forest. One quoted Biswell, speaking "as the mat of debris on the forest floor slowly burned": "Patience is of first importance. The fire must burn downhill, preferably with about a 20 percent slope. You start at the highest point. *You never speed up the fire*" [italics added].[29] An earlier press release was headlined: "UC Forester Sets Fire to Make Woods Safer," and read:

> With more than 60 persons watching him, Dr. Harold Biswell . . . started a fire in Whitaker's Forest, and nobody tried to stop him. The fire burned along through ground cover of pine needles for 20 minutes—and then flickered out on its own—as he knew it would. And the fire did no damage to the redwoods and pines of the forest—something else he knew all the time.
>
> The fire burned properly and safely, watched by National Park Service, U.S. Forest Service, California Division of Forestry, and University of California personnel, and woods farmers from the surrounding area.[30]

"When Harold would hold his field days showing off what his graduate students were doing," Jim Agee recalled, "these Forest Service guys . . . came down and really started eating this stuff up because they could see that they had some support and some tools for putting fire back on the landscape."[31]

Another graduate student, Bruce Kilgore, had studied the impact of the Whitaker's Forest burns on breeding birds to complete his Ph.D. under Berkeley zoology professor A. Starker Leopold. As an undergraduate back in 1951, he had gone on a field trip with one of Leopold's classes to view Dr. Biswell's burn sites at Hoberg's Resort. Kilgore recalled Leopold telling students "that before long fire would be restored to national parks. It seemed a startling and revolutionary idea at the time."[32] Twelve years later the "Leopold Report," prepared for the secretary of interior, finally paved the way for restoration of fire as a natural force in national parks and for Bruce Kilgore to begin a pioneering role in prescribed burning for that agency.

NOTES

HB—Harold Biswell papers, held in the Bancroft Library, University of California, Berkeley (BANC MSS 2002/67 c).

HW—Harold Weaver collection, held at the Forest History Society, Durham, North Carolina.

1. J. Larry Landers, "About E. V. Komarek, Sr.," High Intensity Fire in Wildlands Management, Challenges and Options, in *17th Tall Timbers Fire Ecology Conference. May 18–21*, 1989 (Tallahassee, FL: Tall Timbers Research, Inc., 1991), 4.

2. E. V. Komarek, *A Quest for Ecological Understanding: The Secretary's Review, March 15, 1958–June 30, 1975*, Miscellaneous Publications No. 5 (Tallahasse, FL: Tall Timbers Research Station, 1977), 31, 32.

3. Landers, "About E. V. Komarek, Sr.," 3

4. Ibid., 4.

5. H. H. Biswell, "Research in Wildland Fire Ecology in California," in *Proceedings First Tall Timbers Fire Ecology Conference* (Tallahassee, FL: Tall Timbers Research Institute, 1962), 64.

6. Komarek to Weaver, December 16, 1962, HW, Box 2, Folder I.

7. Weaver to George S. Kephart, December 19, 1962, HW, Box 2, Folder I.

8. Komarek to Weaver, December 28, 1962, HW, Box 2, Folder I.

9. Weaver to Komarek, January 30, 1963, HW, Box 2, Folder I.

10. Komarek to Weaver, February 6, 1963, HW, Box 2, Folder I.

11. Harold Weaver, "Fire and Management Problems in Ponderosa Pine," in *Third Annual Tall Timbers Fire Ecology Conference, April 9, 10, 1964* (Tallahassee, FL: Tall Timbers Research Institute, 1964), 66.

12. Weaver to Biswell, May 12, 1964, HW, Box 2, Folder I.

13. Stoddard to Weaver, May 31, 1964, HW, Box 2, Folder II.
14. H. H. Biswell, "The Role of Fire in Maintaining Forest Wilderness Quality" (Paper presented at the Second Annual California Plant and Soil Conference, California Chapter, American Society of Agronomy, Davis, CA, February 1, 1973), p. 7 of speech notes, HB.
15. Biswell to Weaver, July 16, 1964, HW, Box 2, Folder II.
16. Emanuel Fritz to Biswell, October 15, 1964, HB.
17. Richard J. Hartesveldt, "Fire Ecology of the Giant Sequoias," *Natural History* (December 1964): 12–19.
18. Weaver to Komarek, June 10, 1965, HW, Box 2, Folder III.
19. Komarek to Weaver, June 15, 1965, HW, Box 2, Folder III.
20. Weaver to Paul Keen, June 24, 1966, HW, Box 2, Folder 16.
21. Biswell to Weaver, September 1, 1966, HB.
22. Weaver to Ashley Schiff, August 10, 1962, HW, Box 2, Folder I.
23. Weaver to Komarek, June 13, 1967, HW, Box 2, Folder 16.
24. James Agee, personal communication, January 8, 2001.
25. Harold H. Biswell, Harry R. Kallander, Roy Komarek, Richard J. Vogl, and Harold Weaver, *Ponderosa Fire Management*, Miscellaneous Publication No. 2 (Tallahassee, FL: Tall Timber Research Station, 1973), 4, 7, 44.
26. E. V. Komarek, "Comments on the History of Controlled Burning in the Southern United States," in *Proceedings, 17th Annual Arizona Watershed Symposium*, Arizona Water Commission Report No. 5, Phoenix, September 19, 1973, p. 6.
27. Landers, "About E. V. Komarek, Sr.," 4.
28. E. V. Komarek Fire Ecology Database, online at http://www.talltimbers.org/feco.html/
29. "UC Uses Fire to Protect Sierra Forest," University of California, Berkeley, press release, September 6, 1972, HB.
30. Press release, "UC Forester Sets Fire to Make Woods Safer." Agricultural Extensions Service, University of California, Berkeley, September 26, 1969, HB.
31. James Agee, personal communication. January 8, 2001.
32. Bruce M. Kilgore, "Fire Management in the National Parks: An Overview," Tall Timbers Fire Ecology Conference, October 8–10, 1974. Missoula, Montana, 1976, p. 45.

7. Dog-Hair Thickets in the National Parks

> A great step forward, of benefit to all of society, was made by the National Park Service in 1968. It is interesting that the National Park Service, which is really not a fire control agency, has taken the lead.
>
> Harold Biswell, 1976[1]

Tucked within "Wildlife Management in the National Parks," the report of a committee chaired by A. Starker Leopold, were these words:

> When the forty-niners poured over the Sierra Nevada into California, those that kept diaries spoke almost to a man of the wide-spaced columns of mature trees that grew on the lower western slope in gigantic magnificence. The ground was a grass parkland, in spring-time carpeted with wildflowers. Deer and bears were abundant. *Today much of the west slope is a dog-hair thicket of young pines, white fir, incense cedar, and mature brush—a direct function of over-protection from natural ground fires.* Within the four national parks—Lassen, Yosemite, Sequoia, and Kings Canyon—the thickets are even more impenetrable than elsewhere. Not only is this accumulation of fuel dangerous to the giant sequoias and other mature trees but the animal life is meager, wildflowers are sparse, and to some at least the vegetative tangle is depressing, not uplifting [italics added].[2]

The Leopold Committee recommendations were presented to Secretary of the Interior Stewart Udall, at the twenty-eighth North

American Wildlife and Natural Resources Conference in 1963. They called for a change of focus from simple protection for wildlife to measures that would preserve habitats. Several paragraphs focused on the effects of fire suppression, particularly in California, forty years after that state's light-burning advocates had been stifled. The overall objective for units of the national park system (NPS), the report suggested, should be to recreate

> a reasonable illusion of primitive America . . . using the utmost in skill, judgment and ecologic sensitivity. Of the various methods of manipulating vegetation, the controlled use of fire is the most "natural" and much the cheapest and easiest to apply. Unfortunately, however, forest and chaparral areas that have been completely protected from fire for long periods may require careful advance treatment before even the first experimental blaze is set. Trees and mature brush may have to be cut, piled, and burned before a creeping ground fire can be risked. Once fuel is reduced, periodic burning can be conducted safely and at low expense. On the other hand, some situations may call for a hot burn. On Isle Royale, moose range is created by periodic holocausts that open the forest canopy.[3]

The Leopold Report characterized questions it posed as "of immense concern to the National Park Service." Its recommendations would prove to be immensely important in changing the direction of NPS fire policy. The complete text of the report was published in *American Forests* in April and also in the *Sierra Club Bulletin* of March 1963. Editor Bruce Kilgore introduced it as, "An Inspired Report on Parks."[4]

After the Leopold Report was published, Bruce Kilgore left the Sierra Club and began his Ph.D. program at UC Berkeley under Dr. Leopold. He worked at Whitaker's Forest with Biswell's burn program, where they were cutting, piling, and burning the undergrowth of white fir and incense cedar, with some fire allowed to spread between piles. Kilgore studied the impact of that program on breeding birds from 1964 to 1967.

Leopold sent a letter to Kilgore on September 11:

> Dear Bruce; The day following our meeting at Whittaker Forest, the Advisory Committee on National Parks Research met all day at Sequoia. . . . Among the various recommendations which will be transmitted to Director Hartzog is a very strong feeling on the part of all of us that a Research Biologist should be attached to Sequoia-King's

> Canyon National Park. . . . some control burning on a rather large scale should be undertaken in the Upper King's Basin as soon as practicable. You are an obvious candidate for this Biologist position. I know you have a thesis to write, but . . . I want to ascertain . . . your interest in this regard, and secondly, your availability. I am not in a position to actually make the offer at this time, but it seems like an almost surefire thing if you are interested. I hope that we can work this out. Hope that all is going well with your bird and bug chasing. Will see you soon. All regards.
>
> A. Starker Leopold, Professor of Zoology and Chief Scientist, National Park Service.[5]

Kilgore was hired for the research biologist position at Sequoia and Kings Canyon National Parks (SEKI) in March 1968.

Born in Los Angeles in 1930, Bruce Kilgore grew up in the suburbs—first Montebello, then Glendale. "I am NOT an unbiased person with respect to Sequoia National Park and the giant sequoia trees," Kilgore says. "My folks first took me . . . to see these trees and camp out among them when I was about 5 years old." They traveled from Montebello in a 1930 Model A Ford, leaving at 4 A.M. to avoid hot weather and frequent boiling of their car radiator as they climbed the steep, sixteen-mile grade toward Giant Forest. "We continued to visit Sequoia a number of times as I grew up." He became an Eagle Scout and attended weekend natural history courses at the Los Angeles Museum of Natural History.[6] As an undergraduate at UC Berkeley from 1948 to 1952, Kilgore studied wildlife conservation. In 1963 he returned to Berkeley to work toward his Ph.D. in zoology.

A major turning point for the NPS policy on fire, preceding the Leopold Report, was a wildfire in 1955 on the McGee Ranch west of Kings Canyon. That wildfire rapidly burned more than 13,000 acres of brush and forest and threatened the sequoias at Grant Grove. Harold Biswell and Richard Hartesveldt began pointing to the threat to the Big Trees that kept growing with the suppression of all fires.

PRESERVATION—OBJECTS OR PROCESSES?

The history of fire exclusion efforts in the national parks had paralleled the efforts and policies of the young Forest Service. In March 1885 a special House Committee dealing with the new Yellowstone National Park declared that "the most important duty of the superintendent and assistants in the Park is to protect the forests from fire and ax." The summers of 1889 and 1890 found U.S. Cavalry troops patrolling the park, mainly engaged in extinguishing fires started by careless campers; 61 separate fires were put out by Captain F. A. Boutelle's troops the summer of '89.[7]

The light burning debate in California had only recently been "settled" when, in 1925, Dr. Charles C. Adams of the New York State College of Forestry wrote about "Ecological Conditions in National Forests and in National Parks" for *Scientific Monthly*. Local officials at Sequoia National Park, Adams said, "have strongly advocated 'light burning' of the forest floor, as a means of getting rid of the inflammable litter, including the old fallen trees and stumps. Fires have been a permanent feature of the region, and their complete exclusion will probably in time completely change the appearance of the forest." But Adams felt that no one yet knew enough, or had the financial backing necessary, to successfully practice light burning to maintain virgin forest conditions. Adams was aware of Show and Kotok's research, published in 1924: "Now the results of many years of careful study are available, showing that not only do such fires not really reduce the fire hazard but in fact increase it," while costing more than a program of adequate fire protection. "Every field naturalist knows that to remove the dead and down timber would at the same time destroy the necessary breeding sites and cover for many animals. Thus, from every standpoint 'light burning' should not be practiced in our national parks," Adams concluded.[8]

Yet, actual conditions in the sequoia groves kept convincing officials there on the ground that fire exclusion was a mistake, putting them at odds with the prevalent thinking. Park Superintendent John R. White wrote a letter to the editor of the *Los Angeles Times* on August 29, 1928, pleased that the newspaper had not written colorful accounts of summer wildfires then burning in the San Joaquin Valley west of Sequoia and General Grant National Parks. He described how the forests evolved with fire and were not hurt by it, calling forest fires under primeval conditions, "retail fires." With conditions brought about later by humanity, "wholesale" fires had resulted. "I am indeed glad to see that the *Los Angeles Times* is treating these fires as of so habitual occurrence as scarcely to be news. And that it leaves to lesser papers such scareheads as 'Raging Fire Imperils National Park,' 'Big Trees in Path of Blaze.' They have been in the path of blazes since Ur of the Chaldees reigned in Mesopotamia, yet they still raise their spears or crowns of blue-green foliage to the azure of our Sierra skies."

As CDF historian Raymond Clar noted, "This was especially irritating to the State officials since the State had invested $15,000 and three weeks effort trying to control the particular fire. . . . About a thousand acres of National Park land had been burned and the work performed by that Service was largely negative. Their men simply retreated to the timber line and backfired toward the valley."[9] The NPS administration responded to that incident by hiring a nationwide fire chief, John D. Coffman, who had been Supervisor of the Mendocino National Forest.[10]

The Forest Service fire exclusion philosophy took hold in the NPS, despite a few rangers like White with contrary opinions. NPS Director Arno B. Cammerer wrote luridly about "Outdoor Recreation—Gone With the Flames," for the April 1939 issue of *American Forests*:

> Careless men . . . have destroyed millions of acres of these forests for all time. They have . . . left a glowing campfire to be fanned by the wind from a companionable blaze into a roaring devil of destruction. . . . what they did to the forest was murder. Where there had been beauty, there are now blackened, charred stumps, naked, tortured, twisted tree branches. Where there had been pleasure, there is now pestilence—insects and disease eating out the hearts of the burnt hulks, and erosion destroying the now unprotected earth. Where there had been peace, there is now a new quiet—the quiet of death and stagnation.

An Ansel Adams photograph illustrating that article was captioned: "While the firebrand has not as yet touched this green valley in Yosemite National Park—it can and will whenever vigilance gives way to carelessness."[11]

That caption presented a striking difference to contrary concerns raised just ten years later by a Yosemite National Park forester. Emil Ernst, in May 1949, described vanishing meadows in the valley, due to trees invading the meadows where fires had been excluded. Harold Weaver had seen Ernst's "*Yosemite Nature Notes*" article, but it was a rebuttal article defending protection from fires, written by a park naturalist, Harry Park, for the July 1949 edition, that prompted him to write Ernst:

> Harry C. Park, in his rebuttal, indicates that National Park Policy admits "of only the necessary minimum of artificial management, for

> purposes of administration and for proper public use." He apparently considers that the Indian's intensive use of fire in the Valley constituted a highly artificial method of management. In common with many foresters he overlooks the fact that before interference by the white man fire was a primary ecological factor in the pine forests and that these forests had been conditioned by periodic fires for countless ages.
>
> Present management methods appear highly artificial and, as you suggest, "ill-advised." If fire was of such ecological importance in earlier days, and I believe we have proved conclusively that it was, it should follow that it has its place in management . . . in a National Park as well as to commercial stands, if you are to attempt to preserve the park as nature intended it.[12]

In 1951 Everglades National Park in Florida established two fire lookout stations that soon documented fires ignited naturally by lightning strikes. Park Service biologist William Robertson suggested tests of controlled burning in the sawgrass marshes. "Initially, the Park Service reacted violently," according to historian Ashley Schiff. "The national parks were 'not the place in which to conduct management experiments'."[13] Eventually, permission was granted. (Harold Weaver would tour longleaf pine burning projects in Everglades National Park in 1964).

However, burning in Yosemite Valley was still years off. The first NPS burns outside of the Everglades were in sequoia groves, when Richard Hartesveldt began small experimental burns in 1959. John R. White, the early superintendent, would have been gratified.

Hartesveldt and Harvey's small burns were noticed by the media in 1968. A writer for *Westways* magazine announced: "Smokey Bear has flipped. When last seen, the furry fire fighter had chucked his hat, shovel and badge and was running stark naked through the forest babbling something about rangers setting fires in California's redwood groves." Details of Hartesveldt's research *were* reported positively. "The sunlight streaming down through the 'parklike' open forest in the test plots contrasts sharply with the thick, dark tangle nearby. And in that sunlight 1,000 new *Sequoia giganteas* are living proof that fire is not always the enemy we think."[14]

It was in 1968, not long after Bruce Kilgore was hired at SEKI, that Superintendent John McLaughlin approved an 800-acre prescribed burn in a high elevation, red fir-lodgepole pine forest and, at the same time the first NPS prescribed natural fire in the Middle Fork of the Kings River, above 8,000 feet elevation.

SEQUOIA-KINGS CANYON (SEKI)

Bruce Kilgore recalls flying to the backcountry location with other park staff in a helicopter in May 1968 to look at the proposed burn site with a maverick prescribed burn expert named Harry Schimke. Schimke (yet *another* Harold!) was affiliated with the Pacific Southwest and Range Experiment Station in Berkeley, California, but worked out of Sonora in the Stanislaus National Forest. Schimke "became an early, dogged, and inspired researcher on fuels management" and would eventually be called a "father of modern controlled burns" in a history of the Stanislaus National Forest.[15] "Harry was an expert in fire behavior," Kilgore said.

> He was the guy you called on in those days if you were fighting a wildfire like the McGee fire and you needed someone to light a backfire to try to burn out fuels to stop such a wildfire. But Harry was ahead of his time—and while we in the NPS and Harold Biswell at the UC Berkeley School of Forestry thought highly of Harry and his judgment, not all of his colleagues felt the same about him.
>
> At any rate, that afternoon in the Middle Fork of the Kings, SEKI Chief Ranger Pete Schuft asked Harry whether we were going to be able to control a fire that we started in this forest. We were all novices and depended on the judgment of men like Harry.
>
> I can still remember both Harry Schimke's action and his answer: Harry first grabbed a small shovel, he dug into a nearby large decaying red fir log, stuck his hand into [it] and pulled out a handful of decaying rotten log material, and said, "Hell, Pete, you'll be lucky to get this to burn!" and proceeded to squeeze water out of that handful of down fuel! We gained confidence from people like Harry who had dealt with fire in the woods before. So Pete and his crew of young rangers began prescribed burning in August that summer and burned about 800 acres.
>
> Fire suppression appeared to be of questionable value in this high elevation, near-climax vegetation type unless there was danger to human life or property. So that summer of 1968 we were [also] watching *the first prescribed natural fire allowed to burn in any national park in the nation, the Kennedy Ridge Burn—the first lightning-ignited fire allowed to burn on any public land, to my knowledge* [italics added].[16]

About 85 percent of the two parks, Sequoia and Kings Canyon, ultimately would be designated "natural fire zones." Fires ignited by lightning that were allowed to burn in that zone were, at the time, referred to as "let burns." That terminology changed later, in favor of "prescribed natural fires," because "let burn" gave a mistaken impression of casualness. Such fires were not ignored; park personnel kept close watch by twice-daily fixed-wing flights or other means. However, no suppression action was taken if the fire remained within the zone. As for the location chosen for the very first of such wilderness fires, Tom Nichols, regional fire information officer for the NPS in 2000, commented, "It was so remote, so far away, if it got away, no one would ever know about it."[17]

Before restoring fire to more sensitive and public areas—like giant sequoia groves—a "reduced fuel zone" was prepared with hand-built lines in 1969 along the west boundary of the Redwood Mountain grove. There was clearing around some individual giant sequoias and felling of dead, standing trees. Charged fire hoses were ready on the fire lines. The burn program aimed to reestablish shrubs as part of the wildlife habitat in the groves, foster reproduction of young sequoias, and decrease the unnaturally high fire hazards.[18] A total of one hundred acres were burned between August and November 1969, using a prescription developed by Harry Schimke.

John Bowdler was the fire control officer at SEKI and was in charge of a crew at Redwood Mountain one afternoon in August 1969. His crew was "strip burning." They would burn out a narrow strip along a "scratchline" at the ridgetop that then became a fuel break, creating a widening "black line" of burnt fuels. Then they dropped down, ten to fifteen feet, and, with drip torches, ignited material to burn upslope to the burned zone, where it would go out.

"This gradually becomes a fairly routine process," Bruce Kilgore related.

> As the afternoon wore on, John began noticing that there was no problem if they simply dropped down a bit further and burned out a somewhat wider strip. This seemed to go well for awhile, and the idea came up, why not just finish that last area to be burned as a single strip?
>
> So they did, and guess what? When you drop down THAT much you *can* get into trouble. That strip started benignly enough, but as it moved upslope it picked up speed and intensity. Flame lengths nearly two to three times the height of previous flames leaped into the air beneath some giant sequoias, and John and his crew saw quickly this was not what they had intended to do! They were not in danger of losing the fire . . . BUT . . .
>
> The fire burned so intensely that the lowest branches on some VERY LARGE giant sequoia were killed by heat . . . not by the flames themselves . . . and this was not what the RX burn crews considered

acceptable fire behavior. So this "Bowdler Burn" was pointed to for some time by other burn crews at SEKI as how NOT to do it—even if you want to move more quickly![19]

A concept that Harold Biswell always emphasized—patience, above all, because prescribed burning is not a 9-to-5 job—was reinforced that day. For researchers, it turned out, the unintentionally hot fire of the Bowdler Burn produced better information about seedling success in Sequoias than other burns had, including Dr. Biswell's low-intensity fires at Whitaker's Forest. High intensity broadcast burns, "though they killed lower sequoia branches, ALSO caused the drying of sequoia cones and the deposit of huge numbers of tiny sequoia seeds into nearly ideal mineral soil conditions beneath the mature trees. And they killed out nearly all competing white fir or sugar pine saplings, giving fairly large openings in the forest for ideal growth of young giant sequoia. So what was a RX [prescribed] burn management mistake turned out to be a RX burn research bonanza, and Dr. Harold Biswell and I documented the results of that burn in a *California Agriculture* article published in 1971."[20]

Kilgore explained the NPS philosophy driving the new burn program to a Forest Service audience in Missoula, Montana, in October 1970:

> As a researcher with the National Park Service, I must look at the role of fire in the forest in a somewhat different way from researchers working under other agency philosophies and policies. I feel our present Park Service policy makes the broad philosophical base of our program simpler than that of the Forest Service.
>
> The National Park Service . . . is trying to restore fire to its natural role in forest ecosystems. The simplest way, of course, is to let lightning fires burn. And this is exactly what we have been doing for three years in . . . Kings Canyon National Park.
>
> . . . to date the work at the [Forest Service's] western fire laboratories has been weighed heavily toward fire as it relates to . . . fire control. . . . The work by Harry Schimke until lately of the Pacific Southwest Forest and Range Experiment Station at Sonora is an exception to this. We have attempted, with Schimke's help, to adapt his formulas to the sequoia-mixed conifer forest situation.
>
> . . . the public seems quite ready to accept a reasonable explanation for a shift toward either prescribed burning in our heavily used mixed conifer forest or letting natural fires burn in wilderness areas. The

Bruce Kilgore using a drip torch in a sequoia grove at Redwood Mountain, Kings Canyon National Park, 1969. Courtesy of Bruce Kilgore and the National Park Service.

> biggest problem was within our own agency; I would assume your problem would be the same.
>
> . . . in national parks, our guiding principle is the maintenance of naturalness. And we are finding that whenever and wherever possible, the best way to restore a semblance of native America seems to be to let natural forces run their own course.[21]

John McLaughlin was the SEKI Superintendent who had the courage to authorize the new fire policy. He began his speech at the Tall Timbers Fire Ecology Conference in 1972 with his education—a bachelor's degree in forestry—"since I am regarded in some quarters as somewhat of a fire bug. There is nothing in my background that should lead to this conclusion." Public reaction to the new program had been very positive. "I believe the public attitude with regard to this program is a definite plus," McLaughlin said. "At the same time, I am quite certain that it will be woe to anyone who makes a mistake. In this respect, I suppose one could say we are playing with fire and at this stage there is no column on the score sheet for errors."[22]

YOSEMITE

Resource Manager Robert "Bob" Barbee prepared a 1968 action program to restore surface fires in Yosemite National Park, with input from Harold Biswell and Bruce Kilgore. Six locations in the park were identified for initial treatment: the Mariposa Grove of giant sequoias, the ponderosa pine-bear clover fuels near the Wawona Hotel, the "eleven-mile road" region (in the northwest slope of the park), Yosemite Valley, the Foresta Village (on the western border of the park), and the Tuolumne Grove of giant sequoias and sugar pines. A manpower plan calling for six crews with three members each incorporated recommendations Biswell had been making to agencies for years. The burn team members "should have good judgment and be willing to exercise patience. Since prescribed burning is largely an art and experience is invaluable, these men should be retained over a long period of years. They should become well acquainted with each of the six regions as regards fuels, slopes, wind patterns, behavior of fire under varying fuel and weather conditions, techniques of setting fire, and they must plan carefully so that escapes will not occur. Any fire that kills the over-story canopy over more than an acre or so is a failure for that area." Supplemented with seasonal laborers, the crews would work year-around.[23]

That manpower plan was overly optimistic, however. "None of that ever came to pass," Jan van Wagtendonk recalled. Another of Biswell's graduate students, he would join the Yosemite staff in 1972. "Parks that had burning programs cobbled the crews together from people within their parks. For Yosemite that meant primarily the forestry (tree hazard) crews that worked for the Resources Management Division at that time. Little help was forthcoming from the fire folks although they did conduct small burns from time to time."[24]

Hoping to begin burning in the fall of 1969, Chief Ranger Claude McClain asked Harold Biswell to arrange a time for Barbee and others from the park to visit Whitaker's Forest "and observe what you

Meadow burn in Yosemite Valley in 1970 to remove small trees encroaching on the meadow. The burn was one of the first that reintroduced fire to Yosemite National Park. Courtesy of Jan van Wagtendonk and the USGS.

have been doing, as well as ask a few questions which have been plaguing us and which we feel perhaps you will be able to help us with."[25]

The first prescribed burns in Yosemite National Park *were* set that autumn at the Foresta site and the Mariposa sequoia grove. Bob Barbee told Biswell in a November third letter that the Foresta fire was still burning. "I think we have killed most of the manzanita and a goodly amount of cedar where the fire has crept. It will be interesting to see. We goofed on the P.R. a bit, but now we have a sign down there telling people what is going on." Biswell had, apparently, recently recommended that van Wagtendonk be considered for a biologist position at Yosemite. "Your suggestion, of course, sounds great to me," Barbee wrote. "Next question is, how can it be pulled off, or can it?"[26]

Biswell replied soon, complimenting the work at Mariposa and warning Barbee to closely watch the smoldering fire at Foresta so it did not find a place to turn around and burn uphill. Later that winter he became an official "collaborator" with the park.[27] Also that winter Biswell sent a letter of recommendation to the NPS chief of the Office of Natural Science Studies, suggesting that Jan van Wagtendonk fill the research biologist position at Yosemite. "Jan will be completing his course work for the Ph.D. this spring. When his research has been done, I presume he will be one of the best-trained fire ecologists in the U.S. At least, I hope so." [28]

It would be a couple years before van Wagtendonk was actually hired. Bruce Kilgore left SEKI for San Francisco in 1972, accepting a promotion to associate regional director for Resource Management and Planning in the Western Region office. He was in a position there to have impact on the hiring of NPS scientists. "I knew Jan was just finishing his Ph.D. thesis with Biswell. So, with concurrence from Jan and Harold Biswell, I encouraged [Yosemite NP] to hire Jan—an excellent scientist—before some other agency grabbed him. They did so and the NPS has been the beneficiary ever since."[29]

When Biswell hosted a tour of California sites for Harold Weaver the summer of 1970, they went to NPS burn sites at the Mariposa grove, El Capitan picnic area, El Capitan meadow, and Wawona. Weaver wrote the park superintendent afterward, complimenting the work (and sending along 23 photographic prints). He suggested that, at the Mariposa grove, "a small area of white fir thicket . . . be left uncleared, as a check, and as a contrasting area for the tourists to observe."

The new program in Yosemite was not universally praised. In national parks detractors often were most upset by the "unnaturalness" of fires lit by man, considering that preservation was best accomplished when nature was simply left alone. Ian McMillan, a rancher from Shandon, California, served on the California State Parks Commission; that park system was also considering how to reintroduce fire, something that McMillan adamantly opposed. In a letter to Senator Alan Cranston, he wrote:

> In two recent visits to Yosemite National Park, I have been appalled at a new practice there . . . burning native shrubs and young trees on the amazing pretext of making the park look natural. To me the burned

1890. "Confederate group" of giant sequoias in the Mariposa Grove, Yosemite National Park. Photo by George Reichel. Courtesy of Yosemite National Park library.

> landscape appears to be exactly what it is—an artificial, manmade spectacle, entirely unnatural, incongruous, extremely unpleasant to view, and a flagrant violation of the concept of nature preservation on which the park was founded. . . . In my view the new practice has all the aspects of being another move toward converting Yosemite National Park into a big management project entirely alien to its original and now vital purpose as a place of wilderness where natural conditions are allowed to prevail, *left alone* [emphasis in original].[30]

After the park service hosted a fire meeting at Yosemite in September 1970, an in-house memorandum by Forest Service landscape architect Raymond J. Collins, Jr., was passed along to Yosemite Superintendent Wayne Cone. Collins described scorched trees paralleling the heavily traveled road near Wawona as "atrocious, from the scenic value standpoint," though "satisfactory from the professional forester's viewpoint." As much as twelve feet from the ground,

1970. White fir trees obscured all but the fire-scarred sequoia on the left. Thickets could have produced crown fires that would kill sequoias. Photo by Dan Taylor. Courtesy of Yosemite National Park library.

tree growth was "burned to a crisp, rust-brown color which amplified the meaning of fire and death." Collins felt that the public would not understand the long-term objectives of the burning and was concerned that forestry educators and students "were seeing burning techniques applied without relating the burning to as many other considerations as possible." He suggested that the Forest Service, with its expertise and technical ability, lend assistance on how to use fire with a multidisciplined approach. Collins did close with praise for the Park Service's "attempt to use fire in a constructive way on public lands," adding that, "with some refinement and collaboration . . . this kind of project can have unlimited values."[31]

Yosemite National Park was holding Master Plan hearings in 1971. Harold Biswell presented a statement on the management of the park's forests and meadows. He praised the two-year-old burn program at Yosemite as a start toward "the best and most natural way

2001. A scene similar to the original condition has been restored. Photo by David Carle.

to correct the undesirable situations now existing. . . . [and] to work in harmony with the vital natural process that has been removed—in this case, fire." Biswell urged that there be a sense of urgency about moving the work along rapidly, even if it took emergency funds. "If the stately sugar pines and giant sequoia can be saved and one *big* wildfire prevented, the expenditure of any emergency funds would be well worthwhile."[32]

Ian McMillan sent another lengthy complaint letter, this time directly to Yosemite National Park Superintendent Lynn Thompson in 1972. Philosophical differences most concerned him. McMillan termed the burn program "artificial landscaping," and criticized the Leopold Report as reflecting "an orientation toward game management and habitat manipulation—not park preservation." It constituted an "alien philosophy" that overthrew "a basic park ethic." He added, "I have yet to see valid evidence that the Yosemite Indians and their use of fire were part of a natural condition any more than the fire-setting sheepherders and wood cutters who succeeded them." The scorched trees along the highway near Wawona also bothered McMillan. Trees that had died there had been cut—a "failure of expertise," McMillan declared—and piling them out of sight was "deliberate secrecy [about] these failures."[33]

Superintendent Thompson advised McMillan that *he* had requested that the trees along the Wawona road be cut "as they were considered a hazard and might have fallen onto the highway." His response letter reviewed problems that had come with decades of fire exclusion and explained that a "management philosophy of perpetuating natural processes rather than preserving and protecting objects has evolved." He also mentioned Dr. Jan van Wagtendonk, research biologist on their staff, who, since the initial burns "has developed even more precise data on the optimum conditions under which various vegetative types may be burned."[34]

Jan van Wagtendonk was hired in 1972. He became a specialist in burn prescriptions, refining the original prescriptions worked up for the NPS by Harry Schimke, so that safe but effective fires could be predicted for various fuel conditions.

"One of the first things Doc Biswell had me do when I got to UC Berkeley as a student," Jan van Wagtendonk said, "was to read a stack of literature that primarily had to do with the light-burning controversy from the 1910s. He said, 'Here's all the things I want you to read', just to give me a historical background. That's material I wouldn't even have known about if he hadn't pointed me toward it." Yosemite in 1972 was an exciting time for van Wagtendonk. That year they developed wilderness fire zones and "were picking up bigger and bigger units and we were getting a lot done then. It was gratifying and good timing. The Park Service was ready to move and I was there to help move it."

Looking back twenty-eight years later, van Wagtendonk said, "We were successful against a lot of odds. It is amazing what we were able to do given some of the things thrown in front of us—primarily fire management officers that weren't behind the program and other obstacles. There were a lot of problems between the resource people and the fire people during that period of time. Even as successful as we were, imagine what we could have done had we not had the impediments that we did have."[35]

Philosophy differences, again, were behind most such impediments. Similar problems emerged when one of the National Park Service's founding fathers, former Director Horace M. Albright, learned about

Resting near the end of a successful first day burning in Yosemite National Park after many decades of fire suppression. Harold Biswell and graduate student Jan van Wagtendonk, 1970. Courtesy of Jan van Wagtendonk and the USGS.

a new policy taking effect in July 1972 to allow lightning fires to burn uncontrolled in wilderness areas of Yellowstone National Park. He immediately sent a letter to NPS Director George Hartzog:

> I felt it is important to get my views to you as fast as possible, before something terrible happens that might well seriously affect your status, and undo much of your good work. I refer to the new policy just given out from Yellowstone that lightning (so-called natural) fires are to go uncontrolled in two very large *wilderness areas of the Park*—Mirror Plateau and Two Ocean Plateau—a total of 320,000 acres!!!
>
> The reason given is that fire has a place in an ecosystem and so an experiment will be undertaken in these areas. Of course, you will need only one "experiment" to burn up Yellowstone Park.
>
> The Yellowstone announcement points out that fire has a place in an ecosystem. I don't think so at all. Also, George, remember that

> anything that happens in Yellowstone, even an auto accident, gets into papers everywhere. You could not escape extremely serious criticism if such a fire news was accompanied by the word that it was supposed to burn and would not be controlled. I hope you will contramand this uncontrolled fire policy before the end of this week.
>
> George, I say this with utmost devotion to you: If you do not stop this fire policy, at least for 1972, I'll have to enter the defense of the Yellowstone.[36]

Director Hartzog tried to reassure Albright: "There is little to worry about, I believe, this summer season—in fact, less than usual—both because the weather is thus far cooperating and because I have issued instructions to the field to have an attack force ready to quell any fire likely to damage the park. To come to grips with the concerns of many and resolve the problem for some time to come," Hartzog added that he was asking for a special meeting of Leopold's Natural Science Advisory Committee to provide a policy statement and general management plan concerning fire and fire suppression in Yellowstone and other national parks.[37]

Albright's concerns, at least about media coverage and controversy at Yellowstone, were prophetic, but that would not be apparent for another sixteen years.

Of course, Harold Biswell had endured similar criticisms that paved the way for an easier experience for the generation of prescribed burn pioneers that followed. "I didn't have to suffer that," Jan van Wagtendonk recognized. "He did the suffering for me. Having had him as my mentor helped with much of that, and the time I started at Yosemite was the period that the Park Service had decided that they were going to start using fire more, so it was good timing. It was an exciting time."[38]

Dr. Harold Biswell retired from the University of California in 1973. The university awarded him its Berkeley Citation, their highest award for distinguished achievement, an honor given to very few people. Biswell had worked through whatever suffering there had been during his career, remaining optimistic—and persistent—and, though retiring, must have felt, with van Wagtendonk, that the 1970s were an exciting time. He remained busy as ever, teaching university extension courses and consulting for agencies with prescribed fire programs. For the next eight years, in particular, he assisted California State Parks, developing their department's burn program and designing a state-of-the-art training program for their personnel.

NOTES

HB—Harold Biswell papers held at the Bancroft Library, University of California, Berkeley (BANC MSS 2002/67 c).

HW—Harold Weaver collection, held at the Forest History Society, Durham, North Carolina.

BK—Bruce Kilgore's personal papers.

DPR—California Department of Parks & Recreation, Resource Protection Division files.

1. Harold Biswell, "Some Aspects of Simulated Natural Fires in Vegetation Management," Society for Range Management, Omaha, Nebraska, February 19, 1976. Session introductory comments.

2. A. Starker Leopold, S. A. Cain, C. H. Cottam, Ira N. Gabrielson and T. L. Kimball. "The Leopold Committee Report: Wildlife Management in the National Parks," *American Forests* (April 1963): 34.

3. Ibid., 35.

4. Kilgore had an M.A. in conservation journalism; he edited *National Parks Magazine* from 1957 to 1960; in 1960 he was hired by David Brower to edit *Sierra Club Bulletin* and served there until 1965.

5. Leopold to Kilgore, September 11, 1967, BK.

6. Bruce Kilgore, personal communication. January 24, 2001. At the Museum of Natural History classes he met Mary and Bill Hood. When he began working with the National Parks Service in 1968, the Hoods provided copies of early prefire suppression photos at the Confederate Grove in Yosemite that allowed visual documentation of vegetation and fuel changes that came with fire exclusion.

7. H. Duane Hampton, *How the U.S. Cavalry Saved Our National Parks* (Bloomington: Indiana University Press, 1973), 62, 100.

8. Charles Adams, "Ecological Conditions in National Forests and in National Parks." *The Scientific Monthly* 20 (June 1925): 563.

9. Raymond C. Clar, *California Government and Forestry: From Spanish Days until the Creation of the Department of Natural Resources in 1927* (Sacramento: California State Board of Forestry, 1959): 47.

10. Called the "California National Forest" at that time.

11. Arno B. Cammerer, "Outdoor Recreation—Gone with the Flames," *American Forests* (April 1939): 182, 185.

12. Weaver to Ernst, February 21, 1951, HB. In reference to: Emil F. Ernst, "Vanishing Meadows in Yosemite Valley," *Yosemite Nature Notes* 28, no. 5 (May 1951). Also to: Harry C. Parker, "Has Protection Worked Destruction?" *Yosemite Nature Notes* 28, no. 7 (July 1949).

13. Ashley L. Schiff, "Innovation and Administrative Decision Making: The Conservation of Land Resources," *Administrative Science Quarterly* 11, no. 1 (June 1966): 15.

14. Ron Taylor, "Fire in the Redwoods," *Westways* (August 1968): 36–37.

15. Voices from the Past 29: "Early Stanislaus Recollections," http://www.r5.fs.fed.us/stanislaus/heritage/voices/voices29.htm

16. Bruce Kilgore, personal communication. July 20, 2000. Kilgore added: "An interesting sidelight is the makeup of that first burn crew of young rangers at SEKI: Bill Ehorn, who later was Superintendent of CHIS [Channel Islands] and REDW [Redwood], Andy Ringgold, later Supt. of CACO [Cape Cod] and REDW, Harry Graffe, later Supt. of Zion, and Mack Shaver, who followed Bill Ehorn as Supt. of CHIS."

17. Personal communication at the Fire Conference 2000, San Diego, California. November 29, 2000.

18. Bruce Kilgore, and H. H. Biswell, "Seedling Germination Following Fire in a Giant Sequoia Forest." *California Agriculture* (February 1971): 10.

19. Kilgore, personal communication, July 20, 2000.

20. Ibid. Article reference is to Kilgore and Biswell, "Seedling Germination," 8–10.

21. B. M. Kilgore, "Research Needed for an Action Program of Restoring Fire to Giant Sequoias," *Intermountain Fire Research Council Symposium on "The Role of Fire in the Intermountain West"* Intermountain Fire Research Council, Missoula MT, (1970): 172–180.

22. John S. McLaughlin, "Restoring Fire to the Environment in Sequoia and Kings Canyon National Parks." Tall Timbers Fire Ecology Conference (12), Tallahassee, Florida, June 8–9, 1972, pp. 391, 394.

23. Bob Barbee, "An Action Program to Restore Surface Fires as a *Natural* Ecological Factor in the Forests and Meadows of Yosemite National Park," unpublished manuscript in Yosemite National Park research library.

24. Jan van Wagtendonk, personal correspondence, May 2, 2001.

25. Claude McClain to Biswell, June 25, 1969, HB

26. Barbee to Biswell, November 3, 1969, HB.

27. Biswell to Barbee, November 7, 1969, HB.

28. Biswell to Dr. Robert M. Linn, February 5, 1970, HB.

29. Kilgore, personal communication, January 24, 2001.

30. Ian I. McMillan to Senator Alan Cranston, August 31, 1970. State Parks Commissioner McMillan, the same day, sent a copy of this correspondence to California State Parks ecologist Fred Meyer to explain his opposition to controlled burns as a tool for state parks resource management. In his cover letter to Meyer, McMillan explained his view "that the parks, state and national, . . . advance the concept of preservation [representing] our most promising and useful to approach to [the environmental crisis]. (See Chapter 8), DPR.

31. Raymond J. Collins, Jr., USFS file memorandum, October 7, 1970, HB.

32. Harold H. Biswell, "Statement on the Management of Forest and Meadow Resources in Yosemite National Park." (For Master Plan hearings, Yosemite Park, September 13, 1971), HB.

33. Ian McMillan to Yosemite Superintendent Lynn Thompson, September 20, 1972. The State Parks Commissioner also sent a "carbon copy" of this letter to State Parks Director William Penn Mott, Jr., DPR.

34. Thompson reply to McMillan, December 1, 1972. Carbon copy sent to James P. Tryner, Chief, Resource Management and Protection Division, California State Parks, DPR.

35. Jan van Wagtendonk, personal communication (interview), November 30, 2000.

36. Horace M. Albright to George B. Hartzog, Jr., July 12, 1971, DPR.

37. Hartzog to Albright, July 26, 1972, DPR.

38. van Wagtendonk, personal communication, November 30, 2000.

8. Burning California State Parks

> I became Dr. Biswell's acolyte when the program started. His program made so much sense to me. I had no previous fire experience. Day after day I bombarded him with questions. To work with him was one of the great experiences in my life.
>
> Glenn Walfoort[1]

"On general principles we all are against burning as a tool in park management, and feel that it should be used only as a desperate expedient," State Park Director Newton B. Drury wrote in a 1956 memorandum to Jim Tryner, the supervisor at Calaveras Big Trees State Park. Tryner had asked for permission to burn in the sequoia groves. "Apart from the hazard involved, burning has the disadvantages of depriving the ground of a considerable amount of water retaining duff that might be returned to it, in this case in the form of chips, and of leaving unsightly piles of ashes and charcoal. However, this may be a situation where we cannot avoid some burning. We have a colossal problem to deal with expeditiously. Burning is of course much more rapid and less costly than chipping, and for this reason, at places and times when the disadvantages are minor or can be compensated, burning may be really justified."[2]

Drury authorized limited burning of carefully sorted slash and debris; not logs or stumps, and not twigs and smaller branches that, he felt, would just decay. Material to be burned also had to be moved away from other fuels and particularly away from any green growth.

State parks in California had their beginning in 1864, when President Abraham Lincoln turned over Yosemite Valley and the Mariposa Grove of giant sequoias to the state. An expanded Yosemite National Park eventually took back those lands, but California's diverse and spectacular environment led to the creation of a state park system in 1927. The first park was Big Basin Redwoods near Santa Cruz, protecting coast redwoods (*Sequoia sempervirens*). Calaveras Big Trees was acquired in 1931 to protect two of the northernmost groves of the Sierra Nevada's giant sequoia (*Sequoiadendron giganteum*).

By 1966 the Department of Parks and Recreation (DPR) managed 180 units with 38 million visitors annually.[3] That year, state parks Director Fred L. Jones received a letter from the chief of the East Bay Regional Park District, who had just read an article by Dr. Harold Biswell in *Park Maintenance* magazine. Reducing fuel hazard with Biswell's burning program sounded "quite similar to the East Bay Regional Park District's thinking on forest management," Chief Parry Laird wrote, "and our staff has proposed that we embark on such a program." There had also been recent news about the state's effort to double funding for fire control. Though state parks was not in the fire-fighting business (that was handled on state lands by CDF), Laird wondered "whether . . . rather than developing the fire-fighting force, it might be wise to invest a portion of this money on an increasing basis each year in the procurement and training of personnel to carry out a massive controlled burning program in the forests."[4]

DPR managers had been following Biswell's work and the research in national parks within California. Fred Meyer supervised the DPR Environmental Resources Section. After attending one of the field trips to Whitaker's Forest, he wrote Biswell on June 7, 1968: "I am confident that the results of your work will have important application in various park programs, but particularly at Calaveras Big Trees State Park."

However, the first burn contemplated for a state park was on the central coast at Montana de Oro State Park. Ian McMillan, the rancher who had complained about the Yosemite program in 1970, was a member of State Parks Commission, a citizens advisory board for statewide issues. When he learned of the interest in restoring fire to park ecosystems, McMillan sent copies of his Yosemite correspondence to Director William Penn Mott, Jr., to explain his philosophical concerns.

DIRECTOR MOTT

Mott told McMillan that they seemed to be in philosophical agreement, but were reaching different conclusions on how best to implement a preservation philosophy. "The protection of 'natural' biological entities must recognize the role of man in the ecology of those entities," Mott wrote,

> and must make provision to counteract or otherwise to deal with his influences. "Protection" in today's frame of reference implies "neglect," because too frequently it involves a failure to recognize the dominant role of man as a factor in every biological situation. The role of fire in the maintenance of a natural California landscape is another dramatic example of man's failure to recognize the ecological facts of life. As long as we exclude fire as an environmental factor, which we now do by our fire protection programs almost throughout California and the nation, we must not delude ourselves by the belief that we are preserving natural conditions.[5]

William Penn Mott, Jr., was born October 19, 1909, in New York City. At sixteen he moved to Michigan. He graduated from Michigan State University in 1931 with a degree in landscape architecture. After graduate school at UC Berkeley, Mott became a NPS landscape architect in the western region office. He took over the Oakland City parks in 1945, serving as their superintendent for seventeen years. In 1962 he became director of the East Bay Regional Parks District and doubled their acreage while expanding that district from five to twenty parks.

Mott served as State Parks Director in the administration of California Governor Ronald Reagan from 1967 to 1975. He would later agree to become Director of the NPS during President Reagan's second term (1985 to 1989). Mott seemed a surprising choice for the conservative Reagan, but he was well qualified and known as an inspirational, idea-a-minute leader. Biographer Mary Ellen Butler tied the selection of Mott in 1967 to Reagan's close advisor, Edwin Meese. Meese had grown up in Oakland when Mott was superintendent of parks, and he had worked during the summers after World War II for the East Bay Regional Park District. Also, Mott and Meese were both alumni of UC Berkeley. "Meese thought Mott would be a 'sparkplug,' who would re-energize the state Parks and Recreation department."[6]

Mott's rationale on the use of fire in parks appalled Commissioner McMillan. It was simply "agricultural burning," he told Mott. "Fire as used by man to produce any plant growth different than would otherwise occur must properly be considered an agricultural implement." The management philosophy was "alien to the principles of environmental preservation on which the State Park System was founded." McMillan said he was an *advocate* of natural, lightning-caused fires in parks, but when humans handled the ignition, the result was artificial. He asked Mott to "review and reconsider the entire background of this movement to establish agricultural burning as a new management practice in both the national and now our own state parks."[7] As a rancher, McMillan had experience with burning and had concluded that, even on private ranches, the results were negative because costs exceeded benefits.

Montana de Oro State Park, south of Morro Bay on the central California coast, had recently been added to the park system. It included a 140-acre coastal terrace that had been grazed, with soils depleted by intensive farming. Exotic annual grasses and coastal scrub plants had invaded what once had been native bunchgrass prairie. Such grasslands had once covered one-fourth of California; only a few relict stands still existed in 1966. Fire had been a key factor for over 10,000 years in the evolution and maintenance of the native grasslands.

The park had a citizens advisory committee chaired by Harold Miossi. Mott received a letter from Miossi in late November objecting to the burn concept as, in his view, contrary to the park prospectus, which aimed to protect, retain, and help regenerate natural features. Miossi recognized that the difference of opinion revolved around the definition of "natural."[8]

One of Mott's chief advisors on this subject was Jim Tryner (the supervisor at Calaveras Big Trees in 1956), who was now in Sacramento headquarters as head of the Resource Management and Protection Division. Fred Meyer worked for Tryner in charge of the Environmental Resources Section. They both saw fire restoration as a most important ecosystem management activity for the State Park System. When Tryner presented the plan for burning on the coastal terrace to the park advisory committee, Commissioner McMillan also attended and spoke against the plan. The committee voted sixteen-to-one to oppose the burning.

McMillan then wrote to Director Mott on March 24, 1971, to further clarify the philosophical roots of his opposition: "Even if the natural process in the park was favoring shrubs, to destroy or nullify this process with fire or any other man-used tool, for purposes of 'manipulating' the scenery, would be to violate and abandon the basic ethic of preservation upon which our State Park System was founded. It is this basic ethic, far more than the park landscape, that I am concerned about."[9]

Tryner received copies of all this correspondence. He told Mott that a natural relationship on the marine terrace could not exist in the absence of fire, and, despite McMillan's support for lightning-started fires, they could not rely on such natural occurrences in that location, with its small acreage and issues of public safety. In a May 4, 1971, memorandum Tryner also added that the department's new ecologist, Dr. W. James Barry, was preparing an updated plan.

Jim Barry was the first California state park plant ecologist. He grew up in the Sierra Nevada foothills, went to the University of Nevada at Reno, then received his master's degree in environmental horticulture and Ph.D. in plant ecology from UC Davis. One of the first tasks Barry performed for state parks was to survey the loss of native grasslands in California due to overgrazing and exotic species.

Barry called for a modest burn on just thirty-five acres. The plan was presented to the State Parks Commission for its approval. Barry was told by Commissioner McMillan that, "if he went ahead with his plan, his career with the Department of Parks and Recreation would be over!"[10] Director Mott countered this threat by appointing a committee headed by Commissioner Thomas M. Bonnicksen, who was working on his Ph.D. at the time. Bonnicksen's dissertation topic was on the natural role of fire in the Sierra Nevada redwood groves.

Opponents of the Montana de Oro burn project accused the department of "changeable objectives" when the new plan emerged with new details. A story in the local *San Luis Obispo Telegram-Tribune* on January 22, 1972, was headlined, "Grass Scheme Smolders at Montana de Oro." The emphasis on destruction of exotic species "represents the latest variant of the state's reasoning," the report stated. "When first proposed the major argument was that burning would help restore poppy fields on the terrace, and subsequent arguments emphasized elimination of coyote brush that has intruded into the plain from nearby hillsides." Jim Barry was quoted: "Our idea is to do what it would take nature 1,000 years to do, by speeding up the soil rebuilding process."

An editorial in the *Telegram-Tribune* on September 28, 1971, cited McMillan as an information source; he was vigorously drumming up support for his position. Headed, "Let Nature Try It at Montana de Oro," the editor wrote that nature "should be given time to see what it can do without assistance from impatient men with fire in their eyes. These men contend that fire is as much a part of nature as air, soil, or water and there can't be natural conditions without fire. Perhaps, but the fire, to be part of natural conditions, should be a natural fire. Fire deliberately set is unnatural." A number of letters of complaint were received by Tryner, Mott, and Governor Reagan.

Early in 1972 McMillan requested copies of publications from L. T. Burcham, the CDF forester who had been one of controlled burning's (and Harold Biswell's) critics through the years. McMillan passed the information on to the *Telegram-Tribune*, who wrote that Burcham's reports contradicted the state's rationale for the burn. DPR had concluded that Indians provided one of the ignition sources for fires (and viewed that as part of the natural environment to be restored), but a Burcham article, written more than a decade earlier, argued that Indians had been too primitive and too few and their burning not extensive enough to modify the landscape.[11]

In a cover letter to McMillan, Burcham had qualified the information he sent. Burcham's position was evolving by then, and he advised McMillan that each case was a separate situation and the material was "not intended to present a case against the use of fire in land management."[12] Those qualifications did not make it into the newspaper coverage. Burcham even called the DPR resource managers, concerned that McMillan's use of the material could cause interagency problems. Fred Meyer informed Mott that Burcham "is much interested in what we plan to do, is not opposed, and does not want to cause us difficulty either directly or indirectly."[13]

McMillan continued to encourage a letter-writing campaign and kept sending his own comments. As a state park commissioner, his input was expected, but McMillan's unresponsiveness to the data and arguments frustrated the park managers. Tryner sent a memo to Mott on November 27, 1972:

> Our problem with Commissioner McMillan is quite clear. He completely fails to recognize the threat or disaster which hangs over our forest lands because of our current over-protection from fire. His insistence on applying the term "agricultural burning" to any controlled use of fire as a management tool gives further evidence of his failure to comprehend our objectives. Agricultural burning is done to produce a commodity crop; it involves the conversion of a natural plant com-

munity to something else. Our objective is to restore fire to its natural and aboriginal role in the environment. . . . Commissioner McMillan and those who support this viewpoint simply do not have an understanding of the facts regarding the role of fire on the wild lands of California. For that reason they cannot grasp its essential role as a management tool. We must not allow their rantings to prevent us from establishing meaningful and effective programs for resource management involving the enlightened use of fire.[14]

On June 13, 1973, the long-delayed burning at Montana de Oro began. Fred Meyer decided that he should protect his new staff ecologist from possible reprisal. "They can't hurt me," he told Jim Barry, and he lit the first match. Only seventeen of the planned thirty-five acres were actually burned, as fog rolled in during the afternoon. They were the first seventeen acres of prescribed fire in a California state park.

"It was kind of exciting," Jim Barry recalls. "I didn't think about the match so much. I was a little upset because I'd put all these monitoring plots in and one of the crew came in and set up the kitchen and that was the end of the plot. We probably had a crew of 100 CDF firefighters there to burn 35 acres; it was a little overkill. It was a pretty safe area on the terrace, where you'd not expect the fire to get away, but, of course, at all of the earlier burns there was a tremendous presence by CDF to keep things from getting away."

Ian McMillan, bitter about the outcome, sent "a few sad remarks" to the editor of the San Luis Obispo *Telegram-Tribune* on June 27,

> on the final success of the state Department of Parks and Recreation in its four-year campaign to burn off the natural scenery at Montana de Oro State Park. Why did they do it? Was it simply because there's no commercial turnover in leaving a landscape alone—no jobs, no studies, no grants or appropriations, no expense accounts, no economic growth? Montana de Oro . . . should not be converted into an agricultural experiment station where land-use technicians would be trained to violate the concepts and principles on which the state park system was founded, and where the visiting public would be subjected to the propaganda of the fire cult.

The letter appeared beneath the headline: "Park Burn a Success for Fire Cult."

"Native grasses spread considerably after that burn," Jim Barry recalls. "But we went on to Calaveras and kind of focused on that.

When I wrote the first plan there, I was very nervous about it. There was huge fuel buildup and I couldn't see a way to do this without going in in the early spring when areas came out of the snow."[15]

CALAVERAS BIG TREES

With advice from Harold Biswell, Jim Barry wrote the initial prescription and burn plan for Calaveras Big Trees State Park. The South Grove of giant sequoias lay in a basin draining from the northeast to the southwest. The giant sequoias were, generally, in the bottom of the basin, with incense cedar, white fir, sugar pine, and ponderosa pine progressing up to the ridge. Fire had been excluded from Calaveras for more than one hundred years.

Because Biswell and the NPS had been burning in giant sequoia groves for the prior decade, at Calaveras there was much less opposition, though the DPR received plenty of concerned advice. The Save-the-Redwoods League had been instrumental in acquiring redwood parks for the California park system. Newton Drury, formerly the director of both the state and national parks, was then head of the league. Drury sent a letter on January 9, 1973, for the league regarding "the possibility of introducing 'light burning' at Calaveras." His use of that old term—"light burning"—showed his long tenure in California resource issues. The term was not in general use by then; in fact, it was avoided as carrying too much emotional historical baggage. "We are inclined to be very conservative about the use of fire as a tool," Drury wrote, "and hope that in most cases other methods can be used."[16]

Drury wrote again in June to say that the league urged that fire only be used *outside* the groves. They were opposed to any fire that could potentially destroy dogwood, azalea, or other ground cover and asked that thinning be done very conservatively. "We are not convinced that the somewhat cavalier attitude of the advocates of prescribed burning in the National Park Service should be the policy followed in achieving natural conditions."[17]

Fred Meyer, Gene Thomas (a DPR forester), and ranger Glenn Walfoort, along with Save-the-Redwoods League representatives, toured the Yosemite burn program sites in May 1973. Besides the clear fire hazard reduction benefits, one goal Fred Meyer saw as important was countering carpenter ants, ". . . probably the only insect that can directly bring about the death of a Sierra redwood tree." The ants were tied to the great increase in white firs that came with fire exclusion. "Aphids commonly inhabit the young white firs, and they are cultivated and cultured by the ants to provide the food supply for their colonies. A management burning program will therefore not only protect the integrity of the plant association, but also afford direct protection to the redwoods against insect attack."[18]

As the date to begin approached, CDF was informed, Yosemite's Jan van Wagtendonk was asked for advice, and Harold Biswell was hired as a consultant. There was an administration change before fall burning could begin, however. Governor Jerry Brown replaced Reagan and the new State Parks director was Herbert Rhodes. Mott wrote Rhodes about the new program, recommending that the new director approve funding for the program. He also warned him about the elements of controversy. "There are groups and individuals who feel very strongly that the use of fire as a management tool is wrong, and they will object strenuously to this concept."[19]

In the autumn of 1975 Biswell began by conducting field seminars in the South Grove of giant sequoias. Glenn Walfoort was the park resource ranger at the time. In the days that followed, Biswell and Walfoort studied fuels and topography in the South Grove. They took fuel moisture readings, temperature, humidity, and weather observations. There had been rain in October, but November was dry. About November 20 Biswell told Walfoort that conditions were just right for burning.

On November 22, 1975, Area Manager Bob Stewart gave Glenn Walfoort a nod to go ahead.

> I lit a drip torch and ignited the pine needles along the fire road above the South Grove of giant sequoias in Calaveras Big Trees State Park. The fire backed gently downslope in the predetermined area. The flames were about a foot high, occasionally flaring up where the flashy fuels were concentrated.
>
> Dr. Harold Biswell was watching the fire intently. "See how the flames kill some of the young trees and leave others?" he exclaimed. "Some of the thickets are almost completely eliminated, others just thinned out, and this cedar thicket here was hardly touched. You can see how fire is a natural landscape architect and was one of the dominant forces that shaped these groves and the primeval forest. . . . Looks like our prescription is just right!"[20]

They were broadcast burning in the forest understory, backing down from the north rim. "Almost immediately [DPR and CDF] got

Ranger Glenn Walfoort at Calaveras Big Trees State Park standing by a giant sequoia surrounded by dead fuel, 1975. Courtesy of Harold Biswell, Jr.

nervous and ordered us to keep the fire up on the rim," Walfoort remembers. "This meant that we had to construct a fire line approximately 200 feet below the rim. Before we finished the South Grove burn we put in several miles of needless fire line. In fairness to the Department, this was all very new to them."[21]

Biswell began a series of seminars, inviting other agencies and interested individuals to the site. The weather cooperated all that winter, with only occasional light showers all the way to February. "It was hard, dirty work," Walfoort recalls, "but the results, I thought, were excellent. Prior to the treatment, one could walk within twenty feet of a huge sequoia and not even notice it."[22]

> There was a mixture of incredulity, approval, disapproval, amazement or dismay, depending on individual degree of acceptance. People couldn't believe how much we accomplished. I think, in the beginning, many in the Department thought Biswell would just hold seminars and

> conferences until his contract would expire, then the Department could go back and discuss it for a few more years.
>
> In the following months many people came to inspect the area. I think the chief criticism was that fire got into a specific tree or area. We did our best to save individual specimen trees but occasionally missed one.[23]

Owen M. Bradley and his wife, Adrienne, were board members of the Calaveras Grove Association. They were ecstatic and full of praise in several letters sent to park administrators in the following months:

> Best Christmas Present I ever received. Both Adrienne and I feel a load of relief.[24]
>
> If you know Dr. Harold Biswell, you realize that the minute we arrived he handed us rakes, and immediately we were helping. . . . Dr. Biswell knows fires, has an immense respect for it, and handles it with care and caution. A good man for such a delicate task. Now that a controlled burning program is at last policy for the South Grove, we can sleep better. Since we first saw the South Grove in 1947, the danger of an uncontrolled fire has haunted us. Now we feel that, given time, the South Grove may soon be truly saved. And safe.[25]
>
> We are well pleased with the way the burn program is being planned and conducted. Although we had faith in the plan for the burn program, it was hard not to be apprehensive until we actually spent time on the scene and could see by the hour what was happening, the extreme care and caution used, and the results which were apparent surprisingly soon after the burn took place.[26]

After much discussion approval was given to continue that spring. Permission was given to burn down to, but not into, the sequoia grove. That meant another fire line had to be constructed. By the end of the burn season, mid-May, 740 acres along the South Grove rim and the south-facing slope were burned.

"Fire Protects Calaveras Big Trees From Fire" was the headline in the *Stockton Record* on May 2, 1976. For that publicity Biswell explained how the threat of holocaust fires killing the Big Trees was being reduced and other ecological benefits: "The very shrubs themselves need fire," he said. Deer brush plants that provided much of the diet for deer and other wildlife would resprout when burned, and their seeds would lie dormant for fifty years if not heated by fire. "And the smoke itself is beneficial," Biswell added. Research was sug-

Harold Biswell showing the depth of duff (about 18 inches) that was cleared by a prescribed burn around a large sugar pine at Calaveras Big Trees State Park. Courtesy of California State Parks.

gesting that smoke increased the trees' resistance to western gall rust and helped prevent root rot.

On December 8, 1976, fire was reintroduced to the sequoia grove itself. Walfoort recalled how they

> reduced fuel around all redwoods and large pines. These last steps were very beneficial, but the miles of fire line proved unnecessary. Biswell, understanding the anxiety and pressure put on the Department, did not complain as far as I know.
>
> Working very closely with Professor Biswell, I continued to learn his procedures. In his discussion he emphasized patience, over and over. In the South Grove we slowly backed the fire downslope. As we progressed, the treated area above us gradually widened, increasing our safety. After burns in the grove we often cleaned up a little by stacking and burning. Jan van Wagtendonk considers this "busy work," but Biswell felt it made things more palatable and pleasing to the Department. It definitely reduced criticism.[27]

Glenn Walfoort's fire crews removed 20,000 understory trees at Calaveras in the 1970s. He became the state park system's first prescribed fire expert. "I became Dr. Biswell's acolyte when the program started," he said. "His program made so much sense to me. I had no previous fire experience. Day after day I bombarded him with questions. To work with him was one of the great experiences in my life." Walfoort had worked as a state park ranger only since 1971; the ranger career was a late change for him. Born in St. Paul, Minnesota, in 1923, he served in the Army Air Corp in World War II, flying twenty-nine combat missions over Germany and France. After twenty-three years in the Air Force as a navigator, he worked for United Airlines from 1966 to 1971. Two years into his state park career, he was at Calaveras, where he would retire in 1982.[28]

Biswell turned most of the burning over to Walfoort, who extended burning into other parts of the park after finishing the South Grove in 1979. He was often invited to speak at the local service club luncheons. There were about 3,000 homes adjacent to the boundaries of Calaveras Big Trees State Park sharing a continuous fuel belt. Working in the 6,000-acre park, they were seldom more than a mile from the boundary and occasionally burned close to someone's backyard.

"We initiated a vigorous public relations campaign," Walfoort said. "As it turned out, we had very few complaints." One message he used during his talks was the need to "remind ourselves that European settlers found healthy, beautiful forests . . . which evolved through millennia without the aid of fire trucks or 'borate bombers.' " When he spoke to the Calaveras Big Trees Association in 1978, he shocked a few of the park's supporters. Walfoort told them that, through overprotection, the park had become a "biological slum" due to the buildup of duff and dead vegetation inhibiting germination of sequoias and making for a constant danger from wildfire.[29]

MANAGEMENT ISSUES—EXPANDING THE PROGRAM

"The Department's attitude throughout was understandably ambivalent," Walfoort says. "Jim Barry and a few others seemed to give unqualified approval. Most understood the problems caused by fire exclusion but were troubled by occasional scorching and loss of a big pine. Those who were upset by the appearance of a particular spot had no idea of the huge fuel load prior to the fire. Follow-up fires will be much easier, safer, and with much less smoke if they are done in a timely manner."[30]

The difficulties within the agency originated more from budget and organizational problems. The prescribed fire program was halted temporarily in 1978 with about 400 acres still untreated in the South Grove. The burning work had never been properly budgeted; it was just added on top of an already busy park operation. Others had to carry the load when Walfoort was working on the burns and a four-wheel-drive vehicle was also pulled away from the normal park patrols.

"Friday and Saturday, November 3 and 4, were two very disheartening days for me at Calaveras Big Trees State Park," Biswell wrote in a 1978 letter to Jim Tryner. He had been pushing for funding for a crash program for the burning needed outside the South Grove and ran into administrative realities. "On Saturday morning I found Glenn to be very depressed. His spirit and great interest . . . had been essentially destroyed. I have the very highest regard for him and want to do everything I can to keep him in the burning program. Glenn is now an expert in prescribed burning. He is above anyone I know in California in training and experience in managing prescribed fires."[31]

Despite Biswell and Walfoort's concerns about the Calaveras program, the prescribed burn program *was* eventually funded through the statewide resource management budget after 1980. Meanwhile, Biswell was training state park ecologists and, with his guidance, burner Jason Greenlee, ecologist Jim Barry, and burn crews conducted the first prescribed burning of *coast* redwood forest on May 9, 1978, at Big Basin Redwoods State Park.

Management burns began at Point Lobos State Reserve in 1981. As at Montana de Oro, restoring native grasslands remained a priority for many coastal sites. The burns at Point Lobos enabled large stands of native bunchgrass to reestablish on grasslands dominated by alien annual grasses. A two-acre stand of nonnative black mustard burned with dramatic, thirty-foot flame lengths. Under the same conditions that produced those mustard flames, native grasses flamed less than a foot high. After five years of burning, mustard diminished to just a few individuals. In that same park an understory burn was conducted in 1982 beneath Monterey pines. Thick duff and litter had accumulated since the 1930s, when the pines had invaded a grassland shown on early vegetation maps. The fire fried the cambium layer of the shallow root system of the pines. An ugly die-off of pines followed and public criticism developed. "Although the objective was to restore grasslands to this area, we had planned to do it gradually rather than all at once," Jim Barry said.[32]

Point Mugu State Park in Southern California conducted experimental burns in native grasslands, coastal sage scrub, and coast live oak woodland beginning in 1981. Dr. Richard Vogl (the CSU Los Angeles professor who served on the board of the Tall Timbers Institute) established study plots to determine the best timing for fires that favored native plants.

Harold Biswell conducted prescribed burns in Cuyamaca Rancho State Park in the mountains east of San Diego, in the spring of 1977 and the following two years. A group called the Mountain Defense League sued to stop a burn planned for Cuyamaca in 1982. They felt that the department's environmental documentation was inadequate.

"They were talking about burning the entire forest in a seven-year period," said Byron Lindsley, Jr., spokesman for the group. "We feel that's like burning the village to save it."[33] Biswell wrote Lindsley on October 24, 1982. He saw the problem as a lack of communication. "Of course we are worried that a prescribed fire might escape. It means we must be very careful. However, I am much more worried about a wildfire burning through the entire park during Santa Ana winds when the humidity is extremely low." The group wanted to protect park vegetation, including white firs and incense-cedar. Biswell explained how fire exclusion had led to unnatural favoring of those species that tolerated shade as seedlings. "In restoring fire, we hope to keep the white fir and incense-cedar in their natural place, but certainly not to get rid of it."[34] The lawsuit was dropped when DPR agreed to complete a comprehensive Environmental Impact Report on the burn project.

TRAINING

Biswell, as a special consultant to DPR from 1975 to 1982, had the opportunity to design the type of training program for prescribed burning that he had been advocating for years. His program focused on hands-on prescribed burning and relatively little classroom work. "The most important step in training personnel is to give them field experience under supervision," he had often said. Personnel chosen for the program needed "good judgment, keen observation, high powers of concentration, energy, interest in the work, and above all, patience." Classroom training included two, six-day courses with college credit through the University of California. "The prescribed burner should be an ecologist," Biswell felt. "He should know the fuels and their flammability, the species and their tolerance to fire, and effects of fire on the environment. He should have good training in silviculture and forest management."[35]

Field work in those early years was conducted in Calaveras Big Trees, Mount Diablo, Folsom Lake, Angel Island, and Cuyamaca Rancho State Parks. Because Biswell also was a consultant for Yosemite National Park and San Diego County, he used their sites for additional training. "Doc" called on many experts from various agencies to cover the range of topics (instructors that he had, in many cases, personally introduced to the work). In 1982 they included Jim Agee and Jan van Wagtendonk and others from federal agencies, along with Glenn Walfoort and Jim Barry from the DPR.

Formal training was then augmented by burn experience. DPR required completion of sixty days of supervised field experience in burning following the classroom work for certification to conduct burns on State Park System lands. Though the training program he designed would be modified in later years, Harold Biswell could point with justifiable pride in 1989 to California State Parks System's emphasis on "training of personnel for prescribed burning, more so than any other state or federal agency."[36]

Park ecologist Jim Barry particularly admired how Biswell handled criticism. "He was so effective because it just rolled off of him," Barry said, "and he never gave up. Others did studies of burned areas. *Harold was the guy who went out and lit the torch.*"[37] Barry recalled one particular day in the South Grove at Calaveras. "Biswell was showing some men from CDF what we had been doing. Harold was explaining what we did and he was feeling the duff and, suddenly, said, 'Boy, this is just right; these conditions are just right.' He bent down and struck a match. And these guys from CDF damn near dropped their breakfasts." Barry laughed, remembering. "It was an area we had lit before, but *he just lit a match*. . . . And those CDF guys were old-line firefighters; they had a different vision than they do in that agency today."[38]

In fact, CDF, the state's lead wildland firefighting agency, was experiencing dramatic changes. In 1980 the state passed a law to help foster prescribed burning on private land. CDF would contract for burns with landowners in a cost-share program, and the state covered up to 70 percent of the liability insurance premiums. Liability against escape fires had been a major hurdle all the way back to the light-burning debates of the 1920s.

Back in 1973, before any match had been struck at Montana de Oro, Jim Tryner had received a memo from James Whitehead, a district superintendent in southern California. The U.S. Forest Service had just announced plans for statewide fuel reduction burns in California. Whitehead was worried that, with many agencies starting to reintroduce fire, state parks were moving too slowly. "If the Forest Service and others go into this program with complete abandon, we will probably end up with the only areas with brush and/or other forest cover. . . . As a result of this position . . . we will undoubtedly be accused of maintaining 'nurseries for wildfire'."[39]

California State Parks successfully avoided that stigma, though the 1980s brought major changes across the nation because the U.S. Forest Service, in particular, was ready to replace "fire control" with "total fire management."

NOTES

HB—Harold Biswell papers, held at the Bancroft Library, University of California, Berkeley (BANC MSS 2002/67 c).

DPR—California Department of Parks & Recreation, Resource Protection Division files.

1. Glenn Walfoort, personal communication, April 17, 2001.
2. Newton B. Drury, Chief (Director) California State Parks, to Clyde L. Newlin, April 9, 1956, DPR.
3. Today there are 260 units on 1.3 million acres.
4. J. Parry Laird to Fred L. Jones, February 2, 1966, DPR.
5. William Penn Mott, Jr., to Ian McMillan, October 6, 1970, DPR.
6. Mary Ellen Butler, *Prophet of the Parks* (Ashburn, VA: National Recreation and Park Association, 1999), 85.
7. McMillan to Mott, November 7, 1970, DPR.
8. Harold Miossi to Mott, November 24, 1970, DPR.
9. McMillan to Mott, March 24, 1971, DPR.
10. W. James Barry and R. Wayne Harrison, "Prescribed Burning in the California State Park System," Presented at the Symposium on Fire in California Ecosystems: Integrating Ecology, Prevention, and Management, San

Diego, California, November 17–20, 1997, p. 3 of unpublished document, DPR.

11. L. T. Burcham, *Planned Burning as a Management Practice for California Wildlands*, California Division of Forestry, Sacramento, 1959.

12. Burcham (for L. A. Moran, State Forester) to McMillan, March 8, 1971, DPR.

13. Memoranda, Peter Gaidula to Fred Meyer, March 7, 1972; and Fred Meyer to Mott, May 9, 1972, DPR.

14. Jim Tryner to Mott, November 27, 1972, DPR.

15. Jim Barry, personal communication, November 27, 2000.

16. Newton Drury (on Save-the-Redwoods letterhead) to Mott, January 9, 1973, DPR.

17. Drury to Mott, June 27, 1973, DPR.

18. Fred Meyer, September 17, 1974, DPR.

19. Mott to Herbert Rhodes, April 8, 1975, DPR.

20. Glenn Walfoort, "A Summary of the Prescribed Burning Program of Calaveras Big Trees State Park," unpublished paper, 1979, DPR.

21. Glenn Walfoort, undated letter, March, 2001, personal correspondence.

22. Glenn Walfoort, "The Guidelines for prescribed burning at Calaveras Big Trees State Park," unpublished paper, June 2, 1986, DPR.

23. Glenn Walfoort, undated letter, March 2001, personal correspondence.

24. Owen M. Bradley to Tryner and Meyers, December 19, 1975, DPR.

25. Bradley to Richard Brock, Manpower Utilization Section, State Parks, May 24, 1976, DPR.

26. Bradley to Herbert Rhodes, June 2, 1976, DPR.

27. Walfoort, March 2001, personal correspondence.

28. After Ranger Walfoort retired, resource ecologist Wayne Harrison was assigned to Calaveras to continue the burn program. Working from that park, his responsibilities were soon extended as statewide coordinator of the DPR burn program throughout California.

29. "Big Trees Park Termed 'Biological Slum'." *Calaveras Enterprise*, March 15, 1978.

30. Ibid.

31. Biswell to Tryner, November 13, 1978, DPR.

32. Barry and Harrison, 1997.

33. *San Diego Union*, July 1982.

34. Biswell to Byron Lindsley, Jr., October 24, 1982, HB.

35. Harold Biswell, "Prescribed Fire as a Management Tool," presented at the Symposium on Environmental Consequences of Fire and Fuel Management in Mediterranean Ecosystems, Palo Alto, California, August 1–5, 1977, pp. 2, 5.

36. Harold Biswell, *Prescribed Burning in California Wildlands Vegetation*

Management (1989; reprint, Berkeley: University of California Press, 1999), 112.

37. Jim Barry, interview, November 30, 2000.
38. Ibid.
39. Memorandum, Whitehead to Tryner, April 24, 1973, DPR.

9. National Fire Management

> We were extremely nervous in the early years. We knew that if we blew one fire, we would lose the whole program. So the question was, could we build public support fast enough that when we had the inevitable loss, people would not pull away from the program?
>
> Orville Daniels, Forest Supervisor at the Bitterroot National Forest in 1972[1]

"We're actually protecting trees to death; we're building toward a disaster situation in the forest," said Robert Mutch, research scientist at the Forest Fire Laboratory in Missoula, Montana. Mutch was quoted in a *Los Angeles Times* front page story on February 15, 1972, headlined: "Challenge for Smokey; Fires Aren't All Bad."

"We have to get beyond the bear in our approach to forest fires," William R. Beaufait told the reporter who wrote that story. Beaufait was another researcher at the Missoula laboratory. "Smokey the Bear is grade-school stuff and grade school is where Smokey belongs."

Mutch had prepared the White Cap Fire Management Plan for a 100-square-mile area in the Selway-Bitterroot Wilderness. The diversified "nonsuppression zone" included shrubfields and ponderosa pine savannas at 3,000 feet, lodgepole pine higher up, and at 8,500 feet, alpine larch. If weather and fuel conditions were "in prescription," fires started by lightning would be allowed to burn in all of those habitats. Mutch expected fast-spreading, low-intensity fires in the ponderosa pines, hotter crown fires among the lodgepoles,

Forest Service scientist Robert Mutch on the Selway-Bitterroot Wilderness natural prescribed fire (non-suppression) zone, 1972. Courtesy of Robert Mutch and the USDA Forest Service.

and relatively slow-moving, small fires up in the sub-alpine environments.[2]

Forest Service administrators had not yet signed off fully on the Selway-Bitterroot project when the *Times* article appeared, but official approval was about to begin a transition from total fire suppression toward a new philosophy that would be called "total fire management." The step was no easier for the Forest Service to take across the nation than it had been in the South.

"It's easier going to bed at night knowing that you've worked all day trying to put a fire out than to go to bed knowing that you plan to do nothing about it," Mutch said. "Also, fire fighting—being close to the smoke and the flame—is kind of exciting. There's a resistance among people who have made their careers in fighting fires to adopting the ecological approach." Beaufait agreed with Mutch: "It takes more courage not to fight fires than to fight them . . . but, sooner or later, these forests are going to burn." He compared the new concept, allowing natural fires to burn, to snow rangers who trigger avalanches before they can hurt people.

Orville Daniels, supervisor of Bitterroot National Forest, approved the test within the Selway-Bitterroot Wilderness. "I had clearly in mind in 1971 that we needed to get fire into all of our ecosystems," Daniels recalled, in 2000, "and that the best place to start was wilderness."[3]

"Wilderness" had become an official designation for portions of America's wildlands in 1964. The Wilderness Act specified that those primeval landscapes least affected by man would be "managed so as to preserve natural conditions."[4] Now, fire was to be restored as a natural force, not only in national park wilderness, but for the first time on such lands within a national forest.

Bob Mutch had visited Sequoia National Park in 1968, when fires were reintroduced there, and taken aerial photographs of one of those first burns. The four-year-old fire program at SEKI was described in the same *Los Angeles Times* article and Bruce Kilgore was quoted: "We can arrogantly presume that we can put out a forest fire, but we know that the forest will burn eventually."

Mutch sent Kilgore several versions of that article from syndicated papers and also shared a copy of a memorandum he had just drafted to the head of Fire Control for the Forest Service, Hank DeBruin[5]: Mutch and DeBruin had been having an on-going discussion about how to "do the professional and total job of fire management and still maintain the understanding and support of the fire departments." Mutch recently had addressed that goal as a guest lecturer, speaking to 100 trainees at the Forest Service's "Fire Generalship School." Mutch had concluded the presentation with these thoughts:

> Does a more natural role for fire in wilderness constitute a threat to fire prevention and fire suppression programs? We don't think so—it may even be beneficial. The historical evolution of fire control efforts in this country certainly follows a logical and defensible sequence of events for a young agency charged with the responsibility of managing forest lands. But our fire concepts need to be just as dynamic as the stands that we have stewardship over. The time is now to re-evaluate our protection attitudes so that we are really managing all lands in harmony with . . . biological systems. Skills and knowledge will have to be developed to do the job. In short, it will require professionalism of the highest degree.[6]

Some trainees had seemed very interested and asked that more time be devoted to such topics in the future. Some trainees pursued a different line of reasoning, Mutch noted, considering fires a threat to the resource and human life that should all be controlled. Yet he was gratified to see that 71 percent of the trainees (responding to a true-false test) considered the following statement to be true: "Intensive fire control has actually increased rather than reduced the chances of a very large fire occurring."

THE FRITZ CREEK FIRE

Lightning ignited the Forest Service's first wilderness prescribed fire on August 18, 1972, in a shrubfield at 4,100 feet. That fire burned for four days—covering an area 24-by-24 feet—then went out. It was the only fire within the study area that summer. "We were extremely nervous in the early years," Daniels recalled. "We knew that if we blew one fire, we would lose the whole program. So the question was, could we build public support fast enough that when we had the inevitable loss, people would not pull away from the program?"[7]

The 1973 fire season was much drier. Six fires were ignited that season within the unit. On August 10 lightning started the Fritz Creek fire in ponderosa pine-savanna. Three days later Daniels sent a fire crew to contain the eastern flank to keep fire from spreading out of the management zone. But the crew was also ordered to leave the rest of the fire alone. "I took tremendous heat from the fire community," he said. "They pretty much considered the whole thing foolishness."[8]

On August 15 the Fritz Creek fire *really* tested the commitment to the new program when winds caused "spotting" outside the boundary. "It would have been easy to scuttle the whole program at that time, and few would have blamed them," the editor of *Western Wildlands* later wrote. "It took knowledge, foresight, and courage to see it through the way they did."[9] Firefighters extinguished the 1,600-acre fire outside the management area, but inside the boundaries, again the fire was allowed to burn. Rain put it out after it burned a total of 43 days and meandered over 1,200 acres. The Fritz Creek fire "began to turn the tide with people," Daniels said. But it was a "terrible transition for us all. It just didn't seem right to walk past a hot spot or a stump. Every one of us had grown up fighting fire."[10]

Some of the written comments sent to the Forest Service after the

1973 fires were reprinted in *Western Wildlands*. Reactions were mixed. Thirty letters of support came from citizens, academics, environmental groups, and agencies like the Idaho Fish and Game Department. On the other side, an executive of the Georgia-Pacific timber company wrote to a congressman, calling it an "incredible" decision "to let northern Idaho burn, since," he added, with amazed sarcasm, " 'fire has always been part of the ecology.' " A retired Forest Service employee wrote, "Since it is the avowed purpose of the U.S. Forest Service to burn up the country—as citizens we should unite and take all matches or fire starting material away from all Forest Service people. . . . We can't trust them. We love our Idaho and don't want it burned up." But the episode reminded another retired Forest Service employee, who supported the fire policy, of a proposal made by "Koch (Chief of Timber Management)" in the late 1930s or early 1940s, to let lightning caused fires burn in that same region. "A substantial amount of support was given the idea," when it was presented in a staff meeting. "However, the fear of public reaction against such a revolutionary change . . . killed the Koch proposal."[11]

The recollections of that writer may have related to a 1947 proposal by Chalmer K. Lyman, fire control officer on the Kaniksu National Forest in Northern Idaho. Lyman wrote an article for the journal *Northwest Science*: "Our Choice—A Mild Singe or a Good Scorching." He had conducted a year of study on prescribed burning in the northern Rocky Mountains. Lyman was "not personally alarmed by opposition" to prescribed burning. The Southeast's proponents, he noted, had faced opposition, but were, by then, treating 480,000 acres. He said that only 15,000 acres had been prescribe burned within his region, but, according to Lyman, even in 1947, "most foresters appreciate the possibilities." It took another twenty-five years, but Lyman's belief that "*the facts will lead the way* to more intensive use of fire," was at last being realized.[12]

The Forest Service's Selway-Bitterroot program did not have to stand alone. Fire management studies were underway in 1974 for the Boundary Water Canoe Area in Minnesota, the Teton Wilderness on the Bridger-Teton National Forest, and the Spanish Peaks Primitive Area on the Gallatin. And prescribed burning was being done by the Forest Service in California. In April 1973 the Cleveland National Forest announced plans to burn about 400 acres that year, as part of a statewide effort at fuel modification.[13]

"These 1970s are exciting times for land managers concerned with fire because fire-impoverished landscapes shout for scientific attention," William "Bud" Moore wrote shortly after he retired from directing fire management for the Forest Service's Northern Region. "Much remains to be done before the American people and land management personnel view forest, grassland and shrubland fires as something less evil than Public Enemy Number One. Sometimes," Moore added, "I sense that the public is more ecologically enlightened than most of us old firefighters."[14]

FIRE MANAGEMENT

The Forest Service formally announced its switch from "fire control" to "fire management" at the 1974 Tall Timbers Fire Ecology Conference held in Missoula, Montana. Orville Daniels told the gathering—over 400 land managers and scientists concerned with fire—about the strong commitment needed if "fire management" was to successfully take agencies in a new direction. "The greatest test of commitment occurs when the fire begins," he said, speaking from powerful firsthand experience. Counter-pressures, in particular, could be expected from the fire suppression community. "[Prescribed fire] threatens the hard-won public image that fire is bad, which is the foundation of fire prevention efforts. It also threatens, in a subtle way, reputations of fire suppression teams when they are instructed to monitor and only partially suppress a fire. Past experiences tell them that their reputations are based on putting out fire rapidly and cheaply."[15]

Bruce Kilgore chaired the "Fire Management Section" at that Tall Timbers Conference. His introductory comments reflected on how much change had occurred in just the four years since the last Missoula fire conference. The philosophy then had been, "we can't gamble with as potent a force as fire—either we must have it under full control at all times or shouldn't use it at all." Kilgore saw the need for facts, first of all, supported by "enthusiasm, commitment and guts." Risk would always exist, but "in the long run, fuel accumulates and another manager at a later time, faces an even tougher decision." Finally, Kilgore described as a "trap" the belief that the public would not understand the new direction. Based on his experiences at Sequoia and Kings Canyon, they would be accepting—not blindly—but after programs were explained openly and fully.[16]

Conference presentations covered NPS burn programs then underway at Rocky Mountain, Grand Teton, Yellowstone, Glacier,

Carlsbad Caverns, North Cascades, Everglades, Sequoia/Kings Canyon, and Yosemite National Parks, plus Saguaro National Monument. Since 1968, in those parks, 274 natural fires had been allowed to burn, covering more than 27,000 acres. Active prescribed burning had only been done at 5 parks; 267 burns had treated some 37,000 acres since 1968 (Everglades accounted for 33,000 of those acres, alone).[17]

Also, at the Missoula conference, Kilgore was honored with the National Fire Management Award. Chief of the Forest Service, John McGuire, made the presentation, saying "Not only did [Bruce Kilgore] give us new insights into fire and its place in forest ecology, but he also helped bridge the gap between fire research and the application of research findings to on-the-ground fire management."

A few weeks later *Time* magazine ran an article that gave Kilgore another opportunity to educate the public. A fire had burned 3,500 acres in Grand Teton National Park that summer and fall. "Is the National Park Service concerned? Not really: its new policy . . . is 'Let 'em burn.' " That catchy phrase, with its air of casualness, was not ideal publicity, but the national news magazine did summarize the ecological and hazard reduction benefits. It devoted considerable space, also to negative public reactions. "Mrs. Miles Seeley, a longtime summer resident of nearby Jackson, Wyo., has begun a petition drive to control what she calls the new 'scorched earth policy.' " Others complained about smoke and reduced visibility. "The Park Service's Bruce Kilgore sighs, 'We've got a major problem in explaining our position to the public.' Which suggests," the *Times* writer concluded, "it may be time to fire Smokey the Bear and hire some new symbolic mascot like Sparky the Fireplug."[18]

In 1978 use of fire finally became part of a national "total fire management" policy on *all* federal public lands. The July issue of *American Forests* announced the historic policy change. "In the future, some forest fires which start on National Forest System lands will be used for predetermined beneficial purposes rather than being put out immediately, Assistant Secretary of Agriculture M. Rupert Cutler has announced. 'Now,' Cutler said, 'if a fire starts in an area previously prescribed for burning to improve natural-resource conditions, such as controlling a certain tree disease, the fire may be confined and managed for that purpose.' "[19] Implementation of the new policy began immediately. All National Forests were expected to have their fire-management plans completed by 1983.

That policy announcement allowed Ed Komarek to declare in 1979 that Tall Timbers Research Station's "fight for broad recognition of control burning as a forest management tool has been won. . . . the U.S. Forest Service has changed the name of its Fire Control Division to Fire Management" and there was growing acceptance in the western United States of burning as a tool to prevent holocaust fires.[20]

Researcher Bob Mutch had a leading role within the Forest Service in the years when that policy decision was finalized. He advocated fire, not simply as a "tool," but as a natural force on landscapes. "Using fire simply because it's cheaper, or substituting chain saws and bulldozers for fire to keep smoke out of the air, may be ignoring basic ecological principles governing the structure and function of ecosystems. I have the impression fire is often approached simply as . . . something we turn on or off, like water from a tap, to suit our needs, and not the needs of the ecosystems being managed. Ask this question: is fire a tool like a screwdriver; or is it a 'tool' like temperature, relative humidity, and precipitation?"[21] Mutch used examples like the seeds of redstem ceanothus shrubs, that can remain dormant up to 150 years, waiting for the heat of fire to activate them when a proper seedbed has been prepared by that fire. A bulldozer, used to simply reduce fuel loads, would not serve such fire-dependent species.

As Kilgore would for the NPS, Mutch worked to educate the public and defend his agency's policy, correcting misconceptions and confusion about wilderness prescribed fires. In 1979, 144,000 acres burned in Idaho, most as wildfires that were fought. The "Forest Forum," section of *American Forests* magazine that November, had a letter from a self-described proponent of wilderness who complained that wilderness fires allowed to burn that summer increased the costs of fire fighting, meant losses of timber, and pulled fire fighters away from other fires. Mutch replied to correct the facts: only 31,400 of the total acres noted by the writer had been fires that were allowed to burn for wilderness purposes. "And millions of dollars certainly were *not* expended in suppression action on Wilderness prescription fires in Idaho," Mutch added. The writer of the complaint letter missed the point that prescribed fires should require less manpower and, ideally, produce little or no suppression costs. "Another popular misconception," Mutch pointed out, "is that 31,400 acres of fire results in tree mortality over 31,400 acres. This is definitely not the case." The complaint about lost "timber" ignored, also, the fact that wilderness areas were not open for logging anyway. But Mutch also reminded the writer that, since many tree species in the northern

Rocky Mountains were fire adapted, productivity was actually well served through prescription fires.[22]

The decade that followed the policy change brought a series of publications out of Forest Service regional experiment stations that illustrated the new commitment. Inside the back cover of "Planning for Prescribed Burning in the Inland Northwest" was a caution that Harold Biswell probably would have approved (considering his emphasis on patience): "If the fire is not accomplishing your objectives, modify burning. If it still does not do what you want, STOP BURNING! A prescribed fire burning in fuel that is too wet or too dry wastes resources and can lead to serious problems."[23]

"Planning and Evaluating Prescribed Fires—A Standard Procedure" was published by the Intermountain Forest and Range Experiment Station, in Ogden, Utah, in 1978. By 1979 a Forest Service pamphlet could point to prescribed fire being used on 100,000 acres in National Forests each year in California. "A lot has changed in the last 10 years in what foresters call fire management and some of these changes are not widely known."[24]

The "Boise" Interagency Fire Center, established back in 1965 to pool Bureau of Land Management, Forest Service, and National Weather Service fire and aviation resources, added the National Park Service, Bureau of Indian Affairs, and Fish and Wildlife Service in the 1970s.[25] As one example of the greater coordination between federal agencies, Bruce Kilgore worked for the Forest Service as project leader of the Wilderness Fire Management Research Project from 1981 to 1985 in Missoula, Montana. Kilgore put together a state of knowledge review paper, "The Role of Fire in Wilderness," for the 1985 National Wilderness Research Conference in Fort Collins, Colorado. He left Missoula to return to the NPS in 1985 as their Western Region chief scientist.

A 1983 summary of the evolution of the vanguard SEKI burn program after 15 years looked back at 236 prescribed natural fires (most of them small, totaling 22,000 acres in the wilderness) and 107 prescribed burns that treated over 23,000 acres. Research and field experience brought refinements to burning techniques in the 1980s.

"Prescribed burning has become more a science than an art as predictive capabilities become more accurate." Fire techniques for prescribed burns added spot ignitions, backing fires and night burning as alternatives to strip burning headfires, "allowing fuel characteristics to determine intensity, and recognizing at last the importance of patchiness in Sierra mixed conifer forest dynamics."[26]

BLACK BARK CONTROVERSY

Eric Barnes, from the community of Three Rivers, bordering Sequoia National Park, wrote Senator Alan Cranston in 1985. He was alarmed and outraged by the black char left on Big Trees at the Giant Forest and Garfield groves by prescribed fires and questioned the professionalism and objectives of the entire program.

Bruce Kilgore drafted a reply to Cranston, sent by the Western Region director, and played a key role in gathering a panel of scientists who reviewed the burn program in 1986. "We put together a 7-member panel chaired by Dr. Norm Christensen of Duke University, to consider whether the NPS should change its orientation to favor esthetic considerations over ecological considerations in planning such prescribed burning. We included 3 landscape architects on the panel to give esthetic considerations a fair hearing."[27] The Park Service suspended the prescribed burning program during the deliberation period. Findings and recommendations of the panel were presented early in 1987 to NPS Director William Penn Mott, the former director of California State Parks. The report supported the SEKI fire program, with some minor modifications to minimize unnecessary charring, but ecological values remained paramount, not to be compromised solely for aesthetics. Christensen's panel recommended that two types of prescribed burns be recognized: "restoration fires" to reduce unnatural fuel accumulations and "simulated natural fires"—including fairly intense fires—that would mimic the primeval fire regime. Creating those two different classifications "helped clarify objectives for both the overall program and specific burns."[28] When critics voiced objections to specific results, it would be easier to explain how the goals on a particular site fit within the broad program and philosophy.

Harold Biswell was not directly involved in the government response to Eric Barne's concerns, but corresponded with Barnes during this time. As convinced as ever of the benefits of direct communication and field inspections, Biswell proposed a lunch meeting in Berkeley and invited Barnes to accompany him on a trip to Calaveras Big Trees State Park to see where rangers were cutting understory fir, piling and burning it with debris from broadcast burns done the previous fall. "Late in July or August, I will be going to the South Grove and I hope you can go there at the same time to see some of the fine work they have done and the new reproduction that I understand is doing exceedingly well."

Biswell was eighty-one-years-old in 1986 (retired since 1973), and, clearly, still quite active. His five-page letter explained his early research in the sequoias with an interesting aside: "The last time I saw Whitaker's forest, two years ago, I was discouraged to see that nothing is being done there at present to control an abundance of white fir and incense-cedar seedlings encroaching on the burned-over areas. The whole area we treated is in dire need of a reburn to kill these seedlings." He fully supported the NPS burn program. "I am a strong advocate of the idea of trying to understand nature and working in harmony with it, not so much against it as has been done in the past. This means I am a strong advocate of using fire, one of the chief natural elements governing the composition and structure of forest stands." Regarding black bark, he wrote, "At first I thought [the NPS] burning was too intense. However, I saw later that most of the char on the fibrous bark of giant sequoias soon mellowed, and, furthermore, new reproduction of giant sequoia was far better in the intensely burned areas than in those lightly burned. This caused me to believe that their burning, after all, was not too intense."[29]

Biswell made his own suggestions in this letter, regarding the make-up of the blue-ribbon review panel that Barnes had proposed: Bob Barbee, as chairman (then superintendent of Yellowstone National Park, he had begun the burn program at Yosemite), Bob Mutch (the Forest Service researcher in wilderness fire), Lin Cotton (a private landscape architect who had consulted on Biswell's projects at Whitaker's Forest), Thomas Harvey (the San Jose State University researcher who worked with Hartesveldt at Sequoia), and Glenn Walfoort (retired, by then, from his state park ranger career—Biswell gave high praise to Walfoort's work at Calaveras and understanding of the fire ecology of giant sequoia).

The review panel *actually* chosen by the NPS was, wisely, more independent than Biswell's would have been. None had been actively involved in the development or implementation of the SEKI prescribed fire program.

More than 4 million acres had been incorporated into one of the nation's largest natural fire management zones back in 1982, including wilderness segments of Yellowstone and Grand Teton National Parks and adjacent U.S. Forest Service wildernesses (Teton, Washakie, North Absaroka, and Absaroka-Beartooth). Interagency agreements there allowed natural prescribed fires to cross agency boundaries. That interagency concept was tested in 1988. Bruce Kilgore and Bob Mutch were together in Yellowstone in early September 1988 when fire swept over Old Faithful.

NOTES

HB—Harold Biswell papers, held at the Bancroft Library, University of California, Berkeley (BANC MSS 2002/67 c).

BK—Bruce Kilgore papers.

1. Sherry Devlin, "Check with Reality: Intense Blazes of 2000 May Be Wake-up Call to Return Fire to Forests," *The Missoulian*, Missoula, Montana, online at August 22, 2000. http://www.fs.fed.us/rm/main/pa/newsclips/00_08/082200_reality.html
2. Robert W. Mutch, "I Thought Forest Fires Were Black!" *Western Wildlands* 1, no. 3 (1974): 16–22.
3. Devlin, "Check with Reality," 1 (online text).
4. Wilderness Act of September 3, 1964, Public Law 88–577.
5. Bob Mutch memorandum to Hank DeBruin, March 10, 1972, BK.
6. Ibid.
7. Devlin, "Check with Reality," 2 (online text).
8. Ibid.
9. "About This Issue . . ." *Western Wildlands* 1, no. 3 (1974), editorial introduction by D.J.B.
10. Devlin, "Check with Reality."
11. Mutch, "I Thought Forest Fires," 22.
12. Chalmer K. Lyman, "Our Choice—A Mild Singe or a Good Scorching," *Northwest Science*. 21, no. 3 (1947): 129–133.
13. John Sweeny, "Forest Land Burnings Start Today," *San Diego Union*, April 24, 1973.
14. William R. Moore, "From Fire Control to Fire Management," *Western Wildlands* 1, no. 3 (1974): 12, 13, 15.
15. Orville Daniels, "Fire Management Takes Commitment," in *Proceedings Tall Timbers Fire Ecology Conference and Fire and Land Management Symposium, October 8–10, 1974*, Missoula, Montana, published by Tall Timbers Research Station, Tallahassee, FL, 1976, 164.
16. Bruce Kilgore, "Introduction—Fire Management Section." Tall Timbers Fire Ecology Conference, October 8–10, 1974, Missoula, Montana, published by Tall Timbers Research Station, Tallahasee, FL, 1976, pp. 7–9.
17. Bruce Kilgore, "Fire Management in the National Parks: An Overview." Tall Timbers Fire Ecology Conference, October 8–10, 1974, Missoula, Montana, published by Tall Timbers Research Station, Tallahassee, FL, 1976, pp. 45–57.
18. "Let 'Em Burn." *Time*, October 28, 1974, p. 57.

19. *American Forests* (July 1978): 33.

20. E. V. Komarek, "The Secretary's Review, Quadrennial Report July 1, 1975–June 30, 1979," Tallahassee, Florida, Tall Timbers Research, Inc., 1979, p. 3.

21. Robert W. Mutch, "Understanding Fire as Process and Tool," Adapted from "Fire Management Today: Tradition and Change in the Forest Service," presented at Society of American Foresters National Convention, Washington, D.C., September 28 to October 2, 1975, p. 13.

22. Robert W. Mutch, "More on Wilderness Fires," Letter to Editor, *American Forests* 86 no. 3 (March 1980): 2.

23. USDA, Forest Service, "Planning for Prescribed Burning in the Inland Northwest," Pacific Northwest Forest and Range Experiment Station, 1978.

24. USDA, Forest Service, *Wildfire—Fire Management on the National Forests in California*, Pamphlet 1979–689–182/10 (Washington, D.C.: Government Printing Office, 1979).

25. In early 1993 the Center's name was changed to "National" Interagency Fire Center (NIFC) to more accurately reflect its national mission.

26. Larry, Bancroft, Thomas Nichols, David Parsons, David Graber, Boyd Evison, and Jan van Wagtendonk, "Evolution of the Natural Fire Management Program at Sequoia and Kings Canyon National Parks," paper presented at the Wilderness Fire Symposium, Missoula, Montana, November 15–18, 1983, pp. 174–180.

27. Bruce Kilgore, personal communication, January 24, 2001.

28. David J. Parsons and Jan W. van Wagtendonk, "Fire Research and Management in the Sierra Nevada National Parks," in *Ecosystem Management in the National Parks*, edited by W. L. Halvorson and G. E. Davis (Tucson: University of Arizona Press, 1996), 34. The Christensen panel report was published as: N. L. Christensen, L. Cotton, T. Harvey, R. Martin, J. McBride, P. Rundel, and R. Wakimoto, "Review of Fire Management Program for Sequoia-Mixed Conifer Forests of Yosemite, Sequoia and Kings Canyon National Parks." Final report to National Park Service. Washington, D.C., 1987.

29. Biswell to Eric Barnes, March 4, 1986, HB.

PART III: TO BURN OR NOT TO BURN IS NOT *THE QUESTION*

It is important for us to keep in mind that fire is a natural part of the environment, about as important as rain and sunshine. Very slowly, but eventually, emphasis on prescribed burning will replace that on fire prevention and suppression. When it does, fire suppression will become more effective and much cheaper. There will be a greater awakening to the fact that fire is natural and that we must use it for our own survival.

Harold Biswell, June 23, 1980

10. Yellowstone, 1988

> We would be well advised to retain enough humility to know that nature will not always be controlled despite our best, most carefully planned management.
>
> Yellowstone National Park, 1988[1]

The September 7, 1988, fire that Bruce Kilgore and Bob Mutch experienced at Old Faithful geyser had been ignited by a woodcutter's cigarette on July 22 in the Targhee National Forest. Suppression efforts had immediately begun on the North Fork fire but, pushed by winds and the driest summer since record-keeping began, it burned into Yellowstone National Park and ultimately totaled 373,000 acres.

Yet the North Fork fire was only one of many that summer. The Greater Yellowstone fires of 1988 generated intense political and media focus on what was often called the "let it burn" policy in Yellowstone National Park. Natural prescribed fires were commonly blamed for all the major fires in the park. The consequences of more than a million acres of charred landscape concerned the general public, who relied on inaccurate media reports to evaluate the impacts to a national treasure. Confusion was widespread about the natural role of fire in the forest type that dominated Yellowstone.

Though the massive ecological event provided an unprecedented opportunity for fire ecology researchers in the years that followed, the average acreage burned by the NPS fell from 32,135 acres per year (from 1983 to 1988) to 3,708 acres per year between 1990 and

1994.[2] The national prescribed fire management program was halted, and all fires were fought while politicians, academics, and agency personnel questioned and modified policies and new fire-management plans were mandated. Over twenty national parks began that planning process; three years later, by 1991, just eleven had prescribed natural fire programs back on line.[3]

The summer of 1988 was the driest on record in Yellowstone National Park. Though rainfall in April was 155 percent of normal and in May was 181 percent of normal, practically no rain fell in June, July, or August. That had never happened in the 112-year history of the park. "By late July, moisture content of grasses and small branches in the park reached levels as low as two or three percent, and down trees . . . measured at seven percent (kiln-dried lumber is 12 percent)."[4]

By July 15 it was clear that historic weather patterns and the wetter-than-normal spring were no longer relevant. Lightning strikes early in the summer had started more than twenty fires that burned at first "within prescription." Eleven went out on their own, but the others ran into the altered weather patterns of mid-July. After the fifteenth, no new natural fires were allowed to burn (except in cases where new fires were adjacent to existing fires and clearly going to merge with them). As always, human-caused fires had been fought from their inception. After July 21 *all* fires were fought; on that date the fire perimeters enclosed 17,000 acres. With wildfires growing, Secretary of the Interior Hodel visited on July 27 and reaffirmed that the natural fire program had been suspended.

Suppression tactics were criticized by some for not directly challenging the advancing fronts on all fires. Instead, attacks were against the flanks of fires or focused on protecting structures and lives. The heads of high-intensity fires proved impossible to challenge directly, however. Winds carried embers ahead more than a mile in places and ignited new spot fires. Fire jumped across major features like the Grand Canyon of the Yellowstone River that would normally have confined a fire. Even at night, when cooler air typically raises relative humidity, the fires of 1988 would not "lie down." Many Americans seemed to hold the mistaken belief that modern technology and manpower could control *any* forest fire; they were indignant that the fires were not being stopped. The fire suppression effort in the Greater Yellowstone area became the largest ever undertaken in the United States to that time: 9,000 fire fighters, including army and marine units, more than 100 fire engines, and dozens of helicopters. Yet only a change in the weather would finally snuff the fires.

Media reports were often misleading about the actual extent of the burned landscape within the national park. Some of the confusion arose when fires burning throughout the "Greater Yellowstone Area" were simply referred to as the "Yellowstone fire," suggesting to the public that all of the burned acreage was inside the park. Often stories (and the maps used with them) focused only on the total size of fire perimeters. Though maps provided to reporters by agency personnel included statements that only about half the vegetation within many fire perimeters actually burned, such details were widely ignored.

Ultimately about 1.1 million acres fell inside burn boundaries (in the 2.2 million-acre park), but variability led to patchy fires that actually burned about 440,000 acres (20 percent of the park). Only 60 percent of *that* portion experienced intense crown fires (still, of course, enormous totals).[5] The burned acreage totals dwarfed the combined total of 34,175 acres for all sixteen years since 1972, when the natural prescribed fires policy program was adopted in Yellowstone. The largest of those earlier fires was just 7,400 acres.

Fires that started outside and moved into the park produced half of the burn totals in the Greater Yellowstone area and, once they moved in, became fires that received much of the media attention. The Storm Creek fire began from a lightning strike in the Custer National Forest northeast of the park. When it eventually threatened the Cooke City-Silver Gate area, "it received extended national television coverage and was usually reported as a result of Yellowstone Park's natural fire program."[6]

Not all of the Greater Yellowstone fires that contributed to the regional crisis impacted Yellowstone National Park, however. The Canyon Creek fire was ignited by lightning in Lolo National Forest on June 25. It was within a natural prescribed fire zone and initially allowed to burn. It drowsily smoldered for twenty-six days, until wind and weather changes caused it to "wake up." On August 9 suppression crews and equipment went on the attack when the Canyon Creek fire spotted outside its prescribed boundary. But on September 6 and 7 extreme winds hit the region, and the fire raced across 180,000 acres in just two days. It was not until the weather changed again, when rain and snow arrived on September 10, that the fire was finally controlled.

Orville Daniels, the forest supervisor who had fathered the van-

Bison graze as the "North Fork" fire crowns in trees near Old Faithful in Yellowstone National Park, September 7, 1988. Photo by Robert Mutch, U.S. Forest Service scientist.

guard Forest Service wilderness fire management program, was the forest supervisor for this incident. "Despite the fact that Canyon was managed successfully for sixty-five days," Daniels later said, "once it escaped the wilderness boundary we became the target of much criticism. A few days later the criticism intensified. In a sixteen-hour period, between the sixth and seventh of September, this fire expanded from 68,000 acres to nearly a quarter of a million acres. . . . this violent growth was the result of a surfacing, low-level jet stream—something no one could have predicted."[7]

On August 20 many fires that had seemed to be controlled took a wild wind-driven run. That day became known as "Black Saturday." Fire romped across 250 square miles of national park and national forest lands in one day. And, when fire swept over the Old Faithful geyser region on September 7, media restraint was also swept overboard.

"Old Faithful Will Never Be the Same," was the *Chicago Tribune* headline on September 7. "This is what's left of Yellowstone tonight," a NBC television commentator told evening viewers on September 9, as scenes of black, smoking desolation were shown on screen. The clear implication of the visual images was that *all* of the park was in cinders, though the severity of burn effects shown could be found only on relatively small parts of the fire area.

Conrad Smith, professor of journalism at Ohio State University, later analyzed media coverage of the Yellowstone 1988 fires. Not surprisingly, he learned that news photographers persistently sought flames, the bigger the better, for dramatic visuals. "Although much less than 1 percent of the park was burning at any one time, area shots of flames accounted for 21 percent of the 861 news pictures carried by the three networks."[8]

Inaccurate news coverage about the fire policy may have stirred up much of the anger aimed at the Park Service. Though all fires had been fought after July 21, "NBC implied on September 6 that fires were still being allowed to burn and CBS did so on September 7," Smith found. The *New York Times* story on September 22 focused on the "let burn" policy controversy. Beneath the headline, "Ethic of Protecting Land Fueled Yellowstone's Fires," the page one story declared: "Some wildfires were allowed to burn unattended in Yellowstone Park for nearly three weeks after National Park Service officials were ordered to fight all fires, park officials say." Only on the continuation page, deep within the story, was information that the massive army of fire fighters faced too many fires to allow them to fight on all fire fronts at once. Triage was underway, with top priority the protection of property and lives. Yet the implication in that story and other debates at the time was that a philosophical ethic to permit fire to have "free-range" had been adhered to, beyond reason, by the NPS. That same *Times* story also told readers that fires eventually burned "half the 2.2 million acres of the oldest national park." The implication for American citizens who followed the news was that half of their beloved park had been devastated. And the further implication was that a perverse "let it burn" policy was responsible. The article *did* say that "Even the strongest critics of Park Service tactics say highly unusual weather conditions and extreme winds produced fires that no amount of heavy equipment could have stopped."

That reference to heavy equipment was yet another issue in the news. "Contrary to media reports," the park service clarified, in its own publication about the fires, "bulldozers were in fact used in the park when requested by the firebosses, and fire engines were used regularly off roadways in fire suppression efforts." Standard policies guiding the use of such equipment to minimize resource damage, "were rarely considered a hindrance by the firebosses who ran the operations."[9]

Smith closed his analysis of the journalism with a paragraph that aimed to dispel myths that had been perpetuated in media coverage:

> Yellowstone National Park did not burn down in 1988, although some Americans concluded it had after following press accounts of the fires. The let-it-burn policy of the NPS was not what caused so much to burn, despite media suggestions to the contrary. The fires of August and September could not have been put out with heavy equipment such as bulldozers, contrary to the well-publicized comments of some frontline firefighters and area residents. And fires like these do not ordinarily destroy forests or cause animals to flee in terror, as suggested by network anchors and *Bambi*.[10]

Much criticism that summer targeted NPS Director William Penn Mott, Jr. On September 7 Wyoming Senator Malcolm Wallop called for Mott's resignation, saying, "Let nature take its course has damn near destroyed nature in the greater Yellowstone area."[11]

Mott flew to Yellowstone National Park on August 29, 1988. "There he was . . . a trim, square-shouldered, white-haired man with an infectious smile and crinkly blue eyes behind wire-rimmed glasses, telling the world not to worry. 'These fires are really a predictable part of Nature's grand land-management plan,' he explained to the astounded media representatives. 'In fact, they happen once every several hundred years for a purpose.' "[12]

Mott was then seventy-eight years old. When his former boss, Ronald Reagan, began his second term as president, Mott had agreed to take on the NPS director's job. Their collaboration at the state level had been successful for both men, and James Watt's infamous tenure as secretary of interior had just ended before Mott joined the second administration.

Mott refused to resign, and Reagan did not fire him. Mott remained through the rest of the president's second term. Mott's son, John, remembers that his father's "faith in the importance of prescribed burning and fire in the ecosystem never wavered despite all the bad press and controversy. I never heard him badmouth the naysayers. He always focused on the importance of fire in western eco-

National Parks Director William Penn Mott (*left*) being briefed on the status of 1988 fires in Yellowstone National Park. Courtesy of Yellowstone National Park.

systems and never doubted that NPS was following the correct program."[13]

However, Mott told Broc Stenman, director of the California State Parks Mott Training Center, about "pressures he personally was under in justifying his actions to his superiors in the Reagan administration. It was probably the only time I saw any stress in him," Stenman added.[14]

"Although the media generated a great deal of public sympathy and interest toward Yellowstone," Mott wrote in the November 1988 *Courier*, a park service in-house magazine,

> it also has provided a great deal of misinformation, or maybe I should characterize it more correctly as a lack of information, about what took place there. Yellowstone is *not* a charred wasteland; the fire did burn through areas totaling more than one million acres. But fire doesn't burn evenly, so much of the vegetation within those areas was not touched. We need to make sure the public knows that. We want to encourage people to visit the park, for it now offers the opportunity to see nature at work, to see first-hand the regeneration process.[15]

Those with all sorts of political agendas jumped on the bandwagon of controversy. The *Wall Street Journal*, editorialized about a bill to add land to the National Park system: "With Yellowstone Park a smoke-blackened ruin because of failed federal forest-management policies, one might think it highly embarrassing to propose that still more of America's scenery be trusted to the same bunglers. . . . Some environmentalists—ourselves, for example—are appalled at what happened in Yellowstone. We see little reason to reward mismanagement with more responsibility over national treasures." And the "Wise Use" movement jumped in; Ed Wright, editor at *Our Land* magazine, called the Yellowstone fires "a wake-up call to Western conservatives. We see natural regulation, natural fire, and preservation as a way to close off use of the forests."[16]

The beleaguered Mott, the NPS, and the Forest Service must have been relieved when the *Los Angeles Times*, in a September 13 editorial, spoke out in their defense: "The unwarranted criticism of the Park Service, the U.S. Forest Service and environmental experts has reached a level of misinformed hysteria that is racing out of control as the fires have done." And they may have agreed with M. Rupert Cutler, president of Defenders of Wildlife, who wrote (in the September 11 edition of the *New York Times*), "A firestorm of political opportunism and incendiary rhetoric is racing through the treetops of Congress. It's scorching good science, laying waste to common sense and leaving nothing in its wake but smoke and hot air."

"A TREE IS A TREE . . ."

Mott's boss, Ronald Reagan, was widely credited with saying "If you've seen one redwood, you've seen them all," during his 1966 campaign for governor. His actual words were: "A tree is a tree, how many more do you need to look at?"[17] Preservation of redwoods aside, that comment, "a tree is a tree . . . ," and many reactions to the 1988 wildfires, showed widespread confusion about fire ecology concepts in America.

Not all trees and not all forests are alike, and their relationships to fire form part of that variability. Lodgepole pine forests occupying most of the Yellowstone plateau are adapted to a different fire regime than lower elevation forests such as the ponderosa pines where Harold Weaver and Harold Biswell devoted so much attention.

Like some other fire-adapted species, lodgepole pines have a serotinous cone that remains closed until fire heats resin bonds on the scales, allowing them to release seeds onto ground prepared by fire. But in Yellowstone's lodgepole pine forests that happens only at very long intervals. Plant succession and the cycles of growth and death following a severe fire in those forests gradually produces dead material, but it does not burn readily. Many low-intensity groundfires may creep a short way through the moist understory, but the undergrowth does not carry fires into the tree canopy—at least, not until 200 to 400 years have passed, when extensive mortality finally occurs in the bigger trees. By then understory growth is tall enough to provide a ladder of fuel reaching to that canopy. Large, high-intensity, stand-replacement fires then can occur, when drought and high winds coincide. The last such episode had happened in Yellowstone 280 years earlier. Like the Yellowstone fires of 1988, it probably spread flame across thousands of acres in just a few hours. Such massive events are impossible to control, but since they only come along every few centuries, the risk for people and their property is akin to the long-term risks of living near volcanoes or earthquake faults.[18]

All of this can be confusing for people; it would be much simpler if every forest habitat followed one simple pattern. Stand-replacement crown fires in lodgepole pine—at very long intervals—are natural for that forest type. On the other hand, the late twentieth century prevalence of high-intensity crown fires in the West's ponderosa pine forests were the product of unnatural fuel conditions as frequent low-intensity fires were suppressed for so many decades.

Which may explain why Bruce Kilgore saw the events of 1988 as, above all, "a breakdown in public understanding of the natural role of fire in wildlands, and particularly in our ability to communicate through television, radio, and the press with the public about that role in Yellowstone and elsewhere."[19]

In the next year concerns about local tourism were dispelled when visitation to Yellowstone National Park on Memorial Day, the traditional start of the summer season, was 25 percent greater than the year before. The *San Francisco Examiner* announced, on July 6, 1989: "Worries That Fire Would Cut Tourism Prove Unfounded."

By the end of 1988 the report of an Interagency Fire Management Policy Review Team was released.[20] Ten members from the Department of the Interior and Department of Agriculture issued findings that led to new guidelines for federal agencies. They found that the basic objectives of the fire management program were sound, but policies needed to be refined, strengthened, and reaffirmed. Individual parks and wildernesses needed to reevaluate opportunities to use prescribed burning (by planned ignitions) to complement the wilderness fire programs; hazard fuels should be reduced around developments and adjacent to boundaries. Interagency planning would be improved; training, funding, and research to support the fire management program should all be augmented. One major operational change was a requirement for superintendents to certify daily that adequate resources were available to manage fires as weather and fire behavior changed. Contingency plans had to be in place for the times when regional wildfires produced competition for fire suppression crews and equipment. Another finding was that, in some cases, a perception existed that social and economic impacts on local business and communities had not been considered in 1988. There were instances where the primary message communicated by agencies had continued to be the biological value of fire to vegetation and wildlife, even after fires had been declared wildfires and suppression actions were underway.

A second review panel was assembled by the Greater Yellowstone Coordinating Committee, (representing six national forests and two national parks in the region), headed by Norman Christensen (the Duke University professor who had chaired the 1986 review panel regarding scorched sequoias). That panel brought together ecologists with expertise in natural disturbances. "The only way to eliminate wildland fire is to eliminate wildlands," their report stated. "To extirpate fire completely from a wildland ecosystem is to remove an essential component of that wilderness."[21]

Emotion surrounding the fires of 1988 sometimes contaminated the normally polite debates among academics and agency personnel. The issue of "natural ignitions" versus human-ignited fires remained a contentious philosophical point (as it had been for California Parks Commissioner Ian McMillan, back in 1972–1973). Dr. Thomas Bonnicksen, who had been another commissioner during *that* controversy (on the side of prescribed burning), was now a professor at Texas A&M University. In "Fire Gods and Federal Policy," a 1989 article for *American Forests*, Bonnicksen gave his opinion that the Greater Yellowstone wildfires would not have burned 1.4 million acres if "scientific management that utilized prescribed burning" had been the practice on that landscape,

> especially if vigorous suppression efforts had been undertaken by the Park Service when each of the conflagrations began.

> "Letting nature take its course" has turned the clock back thousands of years to a time when people placed their fate in the hands of mythical gods. Decades of research have brought us to the point where scientific management is feasible, yet today the Park Service is relying instead on Mother Nature or God. In the future, managing a Park or Wilderness will require only that rangers stand on mountaintops making incantations to the Greek god Zeus, asking him to send thunderbolts and fashion a new forest with fire. Who needs science when you believe that the gods are managing your forest?[22]

"Replies from the Fire Gods" was published the following spring. Robert Barbee, the superintendent of Yellowstone National Park, answered Bonnicksen. Barbee had begun the fire program in Yosemite National Park twenty years earlier, and the Yellowstone fires meant he, like Mott, personally bridged the pioneering years and this latest landmark in the history of the nation's fire policy. "Bob took much (unwarranted) flack about those fires, and stood the test well," Bruce Kilgore says. "Other Superintendents with less understanding, background—and backbone—might have caved in."[23] Superintendent Barbee was clearly offended by the tone of challenge in the first article. Bonnicksen revealed "appalling ignorance of Yellowstone fire ecology," Barbee declared, when the professor wrote that the fires of 1988 were not a "natural" event (but the result of mismanagement). Researchers had concluded that the fires very nearly replicated events that had occurred every 200 to 400 years, Barbee said, and pointed to researchers' conclusions that, "Even if the NPS had embarked on aggressive prescribed burning in 1972, at the beginning of the Yellowstone natural fire program, 'the amount of area burned would not have changed significantly'." Barbee saw the conflict as a difference over the broader definition of "wilderness" values. "He [Bonnicksen] advocates a very well-behaved sort of nature, and many wilderness advocates want no such thing. The debate goes on."[24]

Scientists Nathan L. Stephenson, David J. Parson, and Howard T. Nichols, "three front-line fire management officials at Sequoia-Kings Canyon National Parks," further defended the science behind the fire management program. "Prometheus, by the way, had a scatterbrained brother named Epimetheus, which means afterthought. Dr. Bonnicksen is correct in stating that research funds are too precious to spend on fending off criticism. It is our intention to use those funds to make the prescribed-fire program of these parks a model of scientifically guided fire management, in which information needs are anticipated,

"Welcome to West Yellowstone Barbee Que." A jab at Superintendent Bob Barbee during the 1988 fires. Photo by Robert Mutch, U.S. Forest Service scientist.

obtained, and used in the planning process, rather than gathered as an afterthought."

Historian Stephen Pyne, author of many books on humanity's history with fire, also faulted the NPS for relying on the "invisible hand of nature," rather than human calculation. Pyne said their position regarding the inevitability of such high-intensity fire events was "flawed and cynical." When the fires were raging "they were a power no different from hurricanes or earthquakes over which humans had little control. But the park could have altered the fuel structure in advance of the outbreak, and it could have intervened when the fires were small." The historical norm, Pyne maintained, was not untouched landscapes, but "human use of fire everywhere and for every conceivable purpose."[25]

Perhaps because of such criticism, the policy reviewers, following the 1988 fires, *had* concluded that increased prescribed burning should complement wilderness fire management. Bruce Kilgore, who served on the Interagency Fire Management Policy Review Team, made that point, adding, "We need to realize that we implicitly have a 'let-flow' policy for volcanoes, a 'let-shake' policy for earthquakes, and a 'let-blow' policy for hurricanes, that are roughly comparable to our 'let-burn' policy for high-intensity natural fires in wilderness." There was a lesson to be learned, Kilgore felt, "of humility in the face of natural forces over which we often exert little control."[26]

That message was reinforced by several speakers at the Tall Timbers Fire Ecology Conference of 1989. Norman Christensen said: "It is clear that the technology to execute planned-ignition programs in heavy fuels such as the lodgepole pine forests of the Yellowstone Plateau does not exist. Forests and shrublands subject to crown fire are notoriously difficult to ignite within assuredly safe prescriptions [but] virtually impossible to contain or control under conditions typical of when fires 'naturally' occurred."[27]

Scientists Don Despain and William Romme told the gathering:

> We can never know for certain what would have happened if other actions had been taken, of course, but it seems unlikely that earlier suppression would have made much difference. More than 95% of the total area burned in the Greater Yellowstone area burned after July 21, the day on which suppression was ordered on all fires, regardless of origin. Three human-caused fires, which were suppressed from the outset, burned hundreds of thousands of acres despite the best efforts of well-equipped and well-trained fire fighters. Given the burning conditions of 1988, it is highly probable that roughly the same area would have been burned even if total suppression had been applied from the beginning to every fire that started in the greater Yellowstone area.[28]

Orville Daniels spoke eloquently at that conference about the Canyon Creek fire and the future:

> Before the smoke had cleared, we were charged with poor management and bad decisions and, in one case, with not knowing our own environment.
>
> I know better.
>
> I know that, although we must manage carefully and with conscience, we need fire to maintain healthy forests and shape wilderness ecosystems.
>
> The political and social pressure to abandon this program will be intense. I believe, however, it will come in an inconspicuous form—a subtle reluctance to commit, rather than a more adamant, more visible, more vocal opposition that can be more rationally confronted.[29]

"THE GREATEST ECOLOGICAL EVENT IN THE HISTORY OF NATIONAL PARKS"[30]

After 1988, 234 post-fire research projects began in Yellowstone National Park. After ten years researchers gathered at a symposium at Montana State University in Bozeman. "The recovery is certainly proceeding much faster than I would have predicted," Norman Christensen said. The forests were largely replacing themselves, as were grasslands and sagebrush on the park's northern range. Elk and bison herds had only been briefly affected. The fire had directly killed less than 1 percent of the large mammals in the park, but there had been a significant elk die-off the winter of 1988–1989. Two studies had determined that most of the elk had died not because of food lost to fire, but rather from the severity of that winter. Stream ecosystems were being monitored; there had been changes in the physical features and invertebrate life in the streams, but fish were doing fine. There had been concerns that exotic weeds would invade the burned areas, but that turned out to be a minimal change. The patchiness of the fire had pushed vegetation back to a variety of succession stages that was similar to the scale of diversity found before the fire. A surprise to some was that soils had not been sterilized by the intense heat of crown fires. Aspens, a tree that was in general decline across the West, were resprouting vigorously from roots and also sprouting from seed, an uncommon occurrence.[31]

As an "ecological event," the fire and the response by natural systems was a research gold mine. But the administrative restrictions that followed the 1988 Yellowstone fires and the delays in returning the program to many parks concerned Jan van Wagtendonk, the fire scientist at Yosemite. In November 2000, at the National Fire Ecology Conference in San Diego, he told me: "Yosemite has a well-established program, based on science, and most of the managers that come there recognize that and are willing to let the program continue. In other parks it is not as well established and managers are still very skittish about it. They will look for an excuse not to do it. Yellowstone was the first excuse. Up until Yellowstone the programs were growing, the effects were coming on, and after that a lot of people backed off; 'whoa, I don't want to get involved in that!' "[32]

In 1993, at a workshop on NPS fire policy, van Wagtendonk had shared more of his personal thoughts (with a disclaimer that they were not necessarily views endorsed by NPS):

> From an ecological point of view, I am concerned about the restriction that the new fire policies place on fire's natural role. . . . we are being asked to perpetuate natural ecosystem processes while, at the same time, interfering with significant ecological events such as large, intense fires.
>
> A fire that is extinguished leaves unburned fuels that continue to accumulate and cause subsequent fires to burn more intensely. In 1991, for example, the Ill Fire was burning above Yosemite Valley in the Illilouette Creek drainage where, in 1974, Yosemite had undergone its first large prescribed natural fire. The 1974 fire burned some 4,000 acres before we controlled one side because smoke was moving into Yosemite Valley. The former director of the National Park Service was present . . . and he wanted something done about the smoke. Had the 1974 fire been allowed to burn to its natural extent, the 1991 Ill fire would not have been as intense, nor would it have produced as much smoke. In all likelihood, we would not have had to extinguish the fire at considerable expense to taxpayers.[33]

"So what *are* we calling let-burns nowadays?" I asked van Wagtendonk, at the San Diego conference.

"Don't call it let burns!" he exclaimed. "Definitely not! The term is in flux right now, but it's not 'let burn.' It's hard for people to not try to attach some name to it, but there is no name that works. We all understand what 'prescribed natural fire' means, but that term fell into disfavor, for whatever reason. The Park Service—I think it was some guy down in Saguaro—coined the term, 'natural prescribed fire'." He laughed. "It has to do with saying, 'Yes, we know it's a lightning fire, but we're only allowing it to burn with prescribed conditions'."[34] (Recent variations on this terminology are "natural wildland fires" and "wildland fire use.")

Many people would like to see the tool of prescribed burns—with deliberate ignitions—be just a temporary or limited measure for wilderness areas. Those may be the only large landscapes where fires can be allowed to run their natural courses. That philosophical debate hinges on how to reset environments where fire was successfully excluded for decades, and also on the ability to tolerate uncontrollable events like the Yellowstone fires of 1988. However, as Bruce Kilgore said at the 1989 Tall Timbers conference, prescribed burning "may also be permanent if condominiums and other human developments take the place of forest vegetation adjacent to designated wilderness. In high risk areas . . . or where former ignitions or spread in adjacent lands are now blocked by human developments—we'll always need to depend on skillful use of mechanical fuel reduction and prescribed burning."[35]

Kilgore's comments were recognition that across much of America human population growth was shaping the wildfires of the future. Increasingly, the battlefront in the war against nature's fire would be at the edge between urban development and fire-adapted wildlands.

NOTES

1. U.S. National Park Service, "The Yellowstone Fires: A Primer on the 1988 Fire Season," Yosemite National Park, October 1, 1988, p. 17.
2. David J. Parsons and Stephen J. Botti. "Restoration of Fire in National Parks," in *The Use of Fire in Forest Restoration*, General Technical Report INT-GTR-341, June 1996, pp. 29–31. Online: http://www.fs.fed.us/rm/pubs/int_gtr341/gtr341_4.html
3. Rodney Norum, "Natural Fire Management in the National Park Service after 1988," in "Workshop on National Parks Fire Policy: Goals, Perceptions and Reality," published by *Renewable Resources Journal* 11, no. 1 (1993): 18.
4. Ibid., 6.
5. Bruce Kilgore and Miron L. Heinselman, "Fire in Wilderness Ecosystems," in John C. Hendee, George H. Stankey, and Robert C. Lucas, eds., *Wilderness Management* (Golden, CO: North American Press, 1990), 322.
6. U.S. National Park Service, "The Yellowstone Fires," 10.
7. Orville Daniels, "A Forest Supervisor's Perspective on the Prescribed Natural Fire Program," in *Proceedings, 17th Tall Timbers Fire Ecology Conference, High Intensity Fire in Wildlands, May 18–21, 1989* (Tallahassee, FL: Tall Timbers Research Station, 1989), 365.
8. Conrad Smith, *Media and Apocalypse: News Coverage of the Yellowstone Forest Fires*, Exxon Valdez *Oil Spill, and Loma Prieta Earthquake* (Westport, CT: Greenwood Press, 1992), 51.
9. U.S. National Park Service, "The Yellowstone Fires," 9.
10. Smith, *Media and Apocalypse*, 69.
11. Mary Ellen Butler, *Prophet of the Parks* (Ashburn, VA: National Recreation and Park Assn., 1999), 38.
12. Ibid., 4, 5.
13. John Mott, May 13, 2001, personal communication.
14. Broc Stenman, May 14, 2001, personal communication.
15. Quoted in Butler, *Prophet*, 218–219.

16. Both quotes repeated in: Micah Morrison *Fire in Paradise: The Yellowstone Fires and Politics of Environmentalism* (New York: HarperCollins, 1993), 224.

17. Ibid.

18. Don G. Despain and William H. Romme, "Ecology and Management of High-Intensity Fires in Yellowstone National Park," in *Proceedings, 17th Tall Timbers Fire Ecology Conference, May 18–21, 1989, High Intensity Fire in Wildlands, Management Challenges and Options* (Tallahassee, FL: Tall Timbers Research Station, 1989) 43–57.

19. Bruce Kilgore, January 21, 2001, personal communication.

20. U.S. Departments of Agriculture and the Interior, "Recommendations of the Fire Management Policy Review Team," *U.S. Federal Register*, 53, no. 244, December 20, 1988, pp. 51,196–51,205.

21. Norman L. Christensen, James K. Agee, Peter F. Brussard, Jay Hughes, Dennis H. Knight, G. Wayne Minshall, James M. Peek, Stephen J. Pyne, Frederick J. Swanson, Jack Ward Thomas, Stephen Wells, Stephen E. Williams and Henry A. Wright, "Interpreting the Yellowstone Fires of 1988," *BioScience* 39, no. 10 (November 1989): 684.

22. Thomas M. Bonnicksen, "Fire Gods and Federal Policy," *American Forests* (July/August 1989): 14, 16.

23. Bruce Kilgore, January 21, 2001, personal communication.

24. Robert D. Barbee, Nathan L. Stephenson, David J. Parsons, and Howard T. Nichols, "Replies from the Fire Gods," *American Forests* (March/April 1990): 34–35, 70.

25. Stephen Pyne, *World Fire: The Culture of Fire on Earth* (New York: Henry Holt and Co., 1995), 263.

26. Bruce Kilgore, January 21, 2001, personal communication.

27. Norman Christensen, "Wilderness and High Intensity Fire: How Much Is Enough?" in *Proceedings, 17th Tall Timbers Fire Ecology Conference, May 18–21, 1989* (Tallahassee, FL: Tall Timbers Research Station, 1989), 19–20.

28. Despain and Romme, "Ecology and Management," 52–53.

29. Daniels, "A Forest Supervisor's," 365–366.

30. Robert Barbee, "Yellowstone Fires 1988: A Special Supplement to Yellowstone Today," Yellowstone National Park, 1988, p. 1.

31. Yvonne Baskin, "Yellowstone Fires: A Decade Later; Ecological Lessons Learned in the Wake of the Conflagration," *BioScience* (February 1999). Online: http://www.findarticles.com/cf_0/m1042/2_49/53889669

32. Jan W. van Wagtendonk, November 29, 2000, interview.

33. Jan W. van Wagtendonk. "Park Goals and Current Fire Policy," in special report: "Workshop on National Parks Fire Policy: Goals, Perceptions, and Reality," *Renewable Resources Journal* 11, no. 1 (Spring 1993): 19.

34. van Wagtendonk, November 29, 2000, interview.

35. Bruce Kilgore, "Management Options and Policy Directions Concerning High-Intensity Fire: A Fire Policy Panel Discussion," in *Proceedings, 17th Tall Timbers Fire Ecology Conference, May 18–21, 1989* (Tallahassee, FL: Tall Timbers Research Station, 1989), 390.

11. On the Edge

> The hillsides are just the appetizer. Houses are the main course; they provide the bulk of the fuel. The fire loves certain types of neighborhoods—and too many of those types seem eager to oblige.
>
> Bill Crosby, 1992[1]

Within minutes of its start, the fire was out of control. Driven by winds out of the east that gusted up to fifty miles per hour, burning embers ignited the neighborhood landscaping and adjacent homes in the Oakland and Berkeley hills. In just the first hour, 790 homes burned. It was October 20, 1991. The Oakland/Berkeley Hills wildfire, the worst in California's history, killed 25 people, injured 150 others, and destroyed 2,449 houses and 437 apartments and condominium residences. It burned only 1,600 acres, but did an estimated $1.5 billion in damage.

The location of that damage was almost as shocking as its scale. The cities of Oakland and Berkeley are urban environments. Though the fire started on wildland hills covered with grass and chaparral shrubs, it burned into a landscaped "forest." "It was not until the fire that people realized how little the overlay of human civilization—freeways, streets, city boundaries—has to do with the forces of nature," Margaret Sullivan wrote in a history of the fire. "They also realized . . . how all of their separate, isolated, and unique neighborhoods were actually part of the same configuration of ridgelines and hills." Sullivan quoted one baffled victim: "We had sidewalks! The way people talk about the fire area, you would think I was Little Red Riding Hood living in the forest!" But her home was in a narrow canyon, with side-ravines full of chaparral scrub vegetation.

As more people moved into the fast-growing western states, a growing wildland-urban interface complicated the job for wildland firefighters. Few urban residents appreciated the nature of the fire-adapted landscapes that surrounded them. When a wildfire roared into town, the common reaction, Sullivan wrote, was "disbelief that the security of one's established, built community could be so susceptible to the forces of nature."[2]

Congested roads, downed power lines, and blinding smoke limited fire fighters' access to the Oakland Hills in 1991. The approach of flames panicked residents trying to flee; when traffic stopped moving, some left their cars and began running. Their abandoned vehicles added to the congestion on steep, narrow roads. Gas lines ruptured. Water pressure failed. The emergency response was so great that radio airwaves became jammed and telephone systems overloaded, making it impossible to coordinate fire-fighting activities. Collapsed structures and the objects within them slid down the steep hills to further block roads; a new term—"fire debris flow"—would be coined by fire scientists to describe the phenomenon.[3]

> While fire officials labeled the cause of the original fire "suspicious," the reasons for the fire's rapid spread were neither suspicious nor surprising. A five-year drought had dried out overgrown grass, bushes, trees, and shrubs, making them easily ignitable. The parched leaves of closely spaced eucalyptus and Monterey pine trees touched in certain areas and overhung homes in others. Untreated wood shingles were the predominant roof covering for homes in the area. Unprotected wood decks extended out from many of the homes and over sloping terrain that was covered with easily ignitable combustible vegetation. That day unseasonably high temperatures, low relative humidity, and strong winds pervaded the area, further setting the stage for potential disaster.[4]

On November 4 Oakland television station (KTVU) news reporter Bob McKenzie interviewed Harold Biswell to get his comments about the fire the week before. Biswell was eighty-six-years-old, was experiencing heart trouble, and seemed frail. He said that a great risk *still* existed and recommended that remaining residents in the East Bay remove eucalyptus trees, trim vegetation, and burn the cuttings. He predicted that such fires would happen again and could be even worse. "Mr. Biswell's prescription for major disaster goes something like this: another freeze this winter . . . another day with an East wind." As the newsman spoke, the camera panned over rooftops in the Berkeley Hills that were coated with thick layers of pine needles from overhanging trees. "He doesn't expect anyone to listen," the reporter added. "Nobody listened to him 20 years ago either."

Twenty years earlier, in the summer of 1970, a fire in the same hills had destroyed thirty-nine homes. On February 2, 1971, Biswell invited faculty members at the University of California to join him on a tour to look at the wildfire situation. "When A. B. (Tony) Mount of Tasmania, Australia, was here in the summer of 1968," Biswell's invitation said, "he was frightened by what he saw in this area. This person has had wide experience with wildfires in eucalyptus forests in Australia. His thoughts were expressed in his article 'An Australian's Impression of North American Attitudes to Fire,' published in the 1969 Proceedings of the Tall Timbers Fire Ecology Conference, as follows:

> Recently there was much fear expressed (and dispelled) about the San Andreas Fault. Nothing has been said of the far greater threat that exists above the ground in the Berkeley area. Here there is the "ideal" fire situation: a combination of topography, climate, vegetation and houses buried in the vegetation. There is apparently no fear of fire in spite of past fire records, in spite of fuel accumulations under the eucalyptus, up their stems, in their crowns (a continuity of fuel not seen in the same species in Tasmania); in spite of the 'perfect' intermixing of openings with tall dry grass surrounded by heavy forest fuel (both pine and eucalyptus).
>
> Very little imagination is required to foresee a catastrophe in this area. . . . Just how big a catastrophe depends on how soon the public can be made aware of the fuel situation. Without this education it is possible that 500 to 1,000 people will die in a single afternoon within the next five years."[5]

Biswell was able to begin some prescribed burning in the eucalyptus groves of the university campus and the neighboring Tilden Park in 1971 and 1972. While his primary research efforts remained focused on Sierra Nevada pine and sequoia forests, his concern about the Bay Area fire risk took precedence when he spoke at a "Fire in the Environment" symposium in Denver, in May 1972. He titled his talk, "Countdown to Disaster." Later, for an article that reviewed his career and research, in the *Berkeley Campus Report* (December 8, 1972), he convinced the reporter that the Berkeley hills "could set us all on fire. Pieces of windblown eucalyptus bark have been known to set fires 20 miles away."

His sense of urgency deepened that winter after freezing weather killed 80 percent of the eucalyptus trees on 3,000 acres in the East Bay. Biswell estimated that 2 to 3 million big trees were dead, along with many more small ones. Fifty tons of debris per acre waited for an ignition source. That fuel load should be reduced to about five tons per acre, Biswell warned. Otherwise, fire in late summer could produce updrafts that might shower flaming eucalyptus bark on rooftops far from the fire and create a firestorm that could destroy hundreds of homes and people. He publicly called for the dead trees to be cut down and removed—a massive logging endeavor that upset many local residents who loved the trees.[6]

Over a million eucalyptus trees, native to Australia, had been planted in California between 1910 and 1913. They were heavily promoted to landowners as lumber trees, yet the blue gum, gray gum, and red gum imported were actually the least suitable for timber of all the eucalyptus types in Australia. Not mentioned by the promoters, also, was the fact that Australians rarely cut eucalyptus for timber until the trees were 300 years old. Or that the trees, with their shredding bark and resin, were highly flammable.[7]

Eucalyptus promotions went along with the real estate "boosterism" of California early in the century. Southern California, in particular, was touted across the nation and the world for its idyllic, "semi-tropical" climate. Population growth far beyond natural limits was made possible by water aqueducts that connected naturally dry coastal basins to the Sierra Nevada snowpack of northern California and to the Colorado River. More and more homes crowded against chaparral-covered hillsides.

Newcomers to southern California eventually learned about the hot dry "Santa Ana" winds of autumn and about the annual fire season that arrived with those winds. Before the turn of the century, when fire suppression had not yet become official policy, "Older residents said that it was a rarity for a year to go by without at least one disastrous fire in the [San Bernardino] mountains." Angeles National Forest Supervisor Mendenhall in 1930 wrote that annual fires were described by early settlers, but they "were not extensive due to the fact that they ran into older burns and checked themselves." In the chaparral one 1910 observer said fires usually died out on the ridges. "No matter how rapidly it advances up a slope, it subsides at the summit and works slowly down the opposite side. . . . Many cases were seen, however, where fires died entirely on their own on the crests of the ridges."[8]

City of Los Angeles engineer William Mulholland, who designed the aqueduct that brought growth-inducing water 300 miles from the Eastern Sierra, was an early student of chaparral fire ecology. He refused to send water department employees to help fight wildfires in the southern California hills in 1908: "I know that my stand in this matter is not supported by any of the forest rangers, but I have come to this decision in regard to the damage done by brush fires, from observation. If a portion of the water shed burns off each year, then there is always a large majority of the shed covered with a new green growth that will defy any fire. . . . experience has taught me that we cannot prevent them. These brush fires burn until they get through and then they quit."[9]

A forerunner of the Oakland/Berkeley Hills fire of 1991 had occurred in 1923. Sixty-eight years later most East Bay residents had forgotten the Giant Berkeley Fire that leveled fifty city blocks. No lives had been lost, but 624 homes, apartments, businesses, and other structures burned down. A sixty-mile-per-hour wind had blown out of the hills that day.

In 1929, after much of the community of Mill Valley (north of the Golden Gate bridge) burned, "a real estate agent put out fliers urging residents to 'defy the fire god' and rebuild. They did."[10] Bay Area communities like Woodside, Rolling Hills, Bradbury and Orinda *insisted* upon board-and-batten, shake-roofed ranchstyle houses, creating communities of "urban fuel."

The "modern age" of wildland fires invading urban areas began in 1961. Before the Bel Air fire in southern California, rarely had as many as 100 houses been consumed by a wildfire; the Bel Air flames consumed 505 homes.

In thirteen days, from September 22 to October 4, 1970, 722 homes were consumed and thousands of structures damaged by more than 700 wildfires in California. The Malibu Canyon fire destroyed 103 homes. In San Diego County, 382 houses were burned down by the Laguna Fire. "Will such a fire disaster occur again in California?" the CDF asked in *California Aflame!* "Under the same conditions, the answer must be, 'Yes—absolutely!'"[11]

Of course, fire-fighting agencies did not find that acceptable. How to stop California's wildland holocausts remained the subject of considerable debate and research. For awhile attention focused on prospects for "type conversion"—replacing the flammable chaparral shrubs with less combustible plants. "In January 1973, the U.S. Forest Service announced a proposal to transform a half-million acres of Southern California chaparral into grassland through prescribed burning, application of herbicides, and mechanized equipment." Environmental groups were dismayed. The Sierra Club told its members that the likely results would be "destruction of wildlife, loss of watershed holding capacity, replacement of native plants with undesirable alien species, buildup of herbicide levels in and around the project areas, further suburban sprawl, and the waste of thousands of dollars of public money." One of the tools in the list was burning. The Sierra Club accepted controlled use of fire to simulate natural conditions, but not "a program of wholesale defoliation."[12] Losses that would have gone along with such natural habitat conversion became increasingly evident, as a lengthening list of endangered species were identified in California's chaparral ecosystems in the following decades.

When Harold Biswell spoke at a symposium at Stanford University in 1977, he included the provocative statement that, "Consideration should be given to passing a law which would require land managers to keep fuels and fire hazards below a certain, tolerable level. With this, it would be unlawful to have [extreme] fuels."[13]

That prompted a complaint letter from CDF Assistant Chief Clinton B. Phillips. Biswell clarified, as he had done so many times, that he did *not* mean that fire suppression was bad. It *was* needed. However, "under our present practices the wildfire problem can only become worse, but I have been saying this for 25 years. I still believe that the most damaging and most expensive wildfires are still ahead of us."[14]

Confusion and misunderstanding about prescribed fire, as Biswell saw it, still blocked effective solutions to the wildfire problem in 1980. That February the *Journal of Forestry* published an article titled "To Burn or Not to Burn: Fire and Chaparral Management in Southern California." U.S. Forest Service researchers Douglas Leisz and Carl Wilson noted that "every siege of disastrous forest fires in California stimulates demands for an increase in prescribed burning to cut down on the hazardous ground fuels and brush." They suggested a greater focus on alternatives for fuel management, like grazing, use of herbicides, machines to harvest chaparral, fuel-breaks and green belts around urban areas, and vegetation type conversion.[15]

Biswell and others saw their approach as a step backward and even a threat to the growing use of prescribed fire The October issue of the journal ran several letters of reaction. Biswell's summarized the increasing size and costs of recent wildfires, then asked, "Why shouldn't people demand something different in fire control methods?" Though Leisz and Wilson said many prescribed fires were too cool to remove enough fuels, "this is exactly what the prescribed burners want," Biswell said. "They do not want the watersheds completely slicked off, as they are by intense wildfires. The Forest Service has been making excellent progress during the past two or three years in restoring fire to its proper role in wildland ecosystems throughout California. I fear that articles of this sort will hinder that progress. That is not a good prospect."

The Forest Service scientists protested that, "By no means do we discount prescribed burning. . . . We try to point out that chaparral burners walk a tricky tightrope." They saw no single, simple, straightforward solution, their point being that "prescribed fire can be one of the methods—not the only one—of effective management."

Biswell was not alone in his concern about the dampening effect of such arguments on the young prescribed fire programs. Ronald H. Wakimoto, professor in the forestry department at UC Berkeley, took the authors to task for not acknowledging fire as natural and necessary to chaparral ecosystems, "nor do they appear to realize that fires will start whether or not managers set them." And Robert L. Koenigs, of the university's cooperative extension service, pointed to the successful burning of 3,000 acres in San Diego County in the prior two years. "The arguments presented in the article fail to deal with the fact that has been proven over and over again: You cannot stop chaparral from burning; you can only delay the inevitable fire."

Despite that exchange, U.S. Forest Service researchers *were* addressing prescribed fire as a tool in chaparral in the 1980s. The experiment station in Berkeley published *Burning by Prescription in Chaparral* in May 1981 by Lisle R. Green. Experimental and operational burns were conducted in all the southern California national forests in the early 1980s. One prescribed burn in the spring of 1980 "paid off for the San Bernardino National Forest during the disastrous Panorama Fire that occurred during the fall of that year. The eastern flank of the fire was stopped at Mud Flats, where it met the

Mud burn and adjacent fuelbreaks. Evaluations following the fire indicated that an additional 6,000 acres could have been lost if the fire had not been halted in the Mud Flats area."[16]

California had 10 million acres of chaparral shrublands that were "fire-dependent." Researchers knew that fire was necessary to keep the chaparral ecosystems vigorous and productive. After a fire many shrubs could resprout from burned stumps. Others had seeds that would only germinate after the intense heat of a fire. The guarantee that fire would return every few decades was assured by flammable resins, oils, and waxes produced most heavily by the plants during the driest seasons. Dead wood remained attached to living shrubs instead of falling to the ground and decomposing. After thirty years shrub fields might have as much dead as living wood. With such traits chaparral seemed designed to burn in a cycle of intense fires that assured its regeneration.

In 1981 the State of California began a program of contracting with private landowners for CDF to conduct prescribed burns. Costs were shared; generally landowners paid 20 to 30 percent and, most significantly, the state covered most of the liability costs. With that assurance the hope was that increased acreage might be burned. According to private fire technology consultant Joe Rawitzer, actually "it set burning in reverse." Back in the 1950s, over 100,000 acres were burned under permits each year. By the 1970s the average had dropped in the face of air quality and liability issues and an increase in the use of herbicides for clearing land. The average burned annually between 1989 and 1999 was 36,455 acres.[17] That was far too little to keep up with the acreage that needed regular treatment. Motives of the legislative sponsors were pure, Rawitzer felt, but "they institutionalized burning; they said, come one, come all, we'll do it for you. Yet the numbers speak for themselves. What they did is create a disincentive for private owners to burn. If you do it yourself and it gets away, [CDF is] going to come and put it out and send you a bill for what it costs to put it out and if it burns off your property, you'll probably be sued."[18]

The CDF program may not yet be as large as needed, yet the goal of addressing liability remained a key issue for prescribed burners across the nation. In Florida far more burning was successfully carried out each year, perhaps because that state's Division of Forestry trained and certified prescribed burn managers—including private individuals who completed their required training. So long as burns were conducted in accordance with Florida guidelines and law, certified burners were not liable for damage or injury caused by their fires or resulting smoke. The Florida certification program took effect in 1988, and the civil liability coverage was added in 1990. (By the year 2000, Florida's private and public burns totaled around 3 million acres every year).[19]

In California other local and regional agencies did move ahead with their own programs. Los Angeles County Fire Department burned over 6,000 acres between 1980 and 1984. Public involvement in that heavily populated county was essential. All of Los Angeles County's prescribed burns at the "urban interface" were required to have a water-dropping helicopter on the scene. Complaints were sometimes received regarding ashes in swimming pools and smoke, but, as the Chaparral Resource Management newsletter reported, "by and large, public response is most positive *when the wildfire alternative is presented*" [italics added].[20] Eleven major wildfires occurred between 1983 and 1988 that "stopped dead where burns had been completed."[21] Those wildfires occurred under high fire risk conditions—dry, hot, and windy.

When satellite imaging of the earth became available, Dr. Richard Minnich used the new tool to map wildfire "mosaics,"—the boundaries of historic fires—both in southern California and south of the international border in Mexico's Baja California. There had been no effort at fire suppression in Baja until about 1960 and no effective suppression even after that time. So Baja offered a "living museum" of the conditions in southern California as it used to be. Minnich's field-proofed maps provided striking visual proof that hundreds of small, low-intensity lightning fires continued to start in Baja each year during early summer thunderstorms. They would go out when they reached the limits of flammable fuel. North of the border, however, in California, there were fewer fires, but more of them were enormous. The line marking the international border was strikingly apparent on fire history maps.

Minnich sent Biswell a copy of his article, "Fire Mosaics in Southern California and Northern Baja California," in 1983. On the title page he penciled a personal note: "It really works; fires run into previous burns and lay down. There is no fire event [in Baja] greater than 10,000 acres over the entire period. Most fires are 1,000 to 3,000 [acres]. Thank you for your interest."[22]

Evidence that prescribed burning could produce barriers to knock down large wildfires was accumulating. A 500-acre wildfire in Orange County, Florida, in February 1991 was pushed by winds into 1,250-

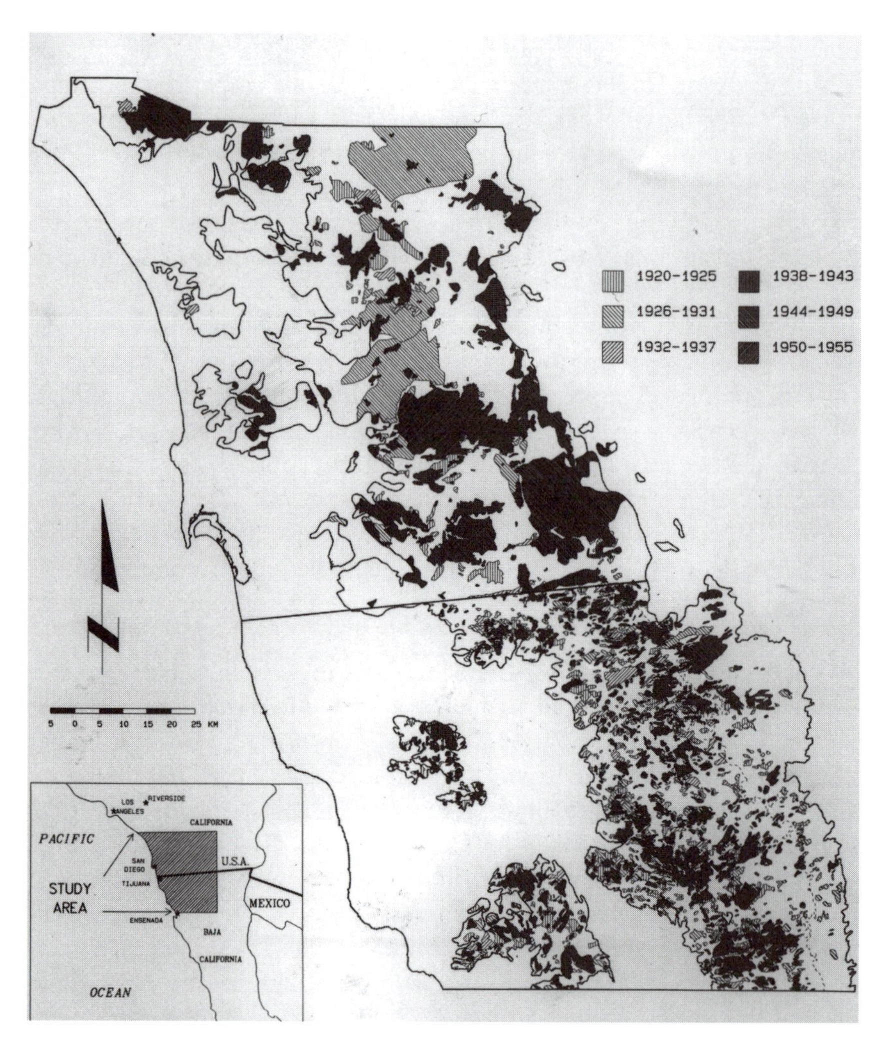

Fire perimeters from 1920–1955 above and below the California/Mexican border. A few enormous fires north of the border compared to many small ones in Mexico; the difference is due to fire suppression in the United States. Map by Dr. Richard A. Minnich, *International Journal of Wildland Fire* 7.3 (1997): 229. Used by permission.

acres that had been prescribe burned in December. "Less than 0.1 acres was burned by the wildfire in the area previously burned by prescribed fire. The fire was spotting 300 to 400 feet ahead of the fire front. If the wildfire had not run into the prescribed burn . . . the fire would easily have tripled in size before the next defensible barrier."[23] The weather and fuel conditions were similar to those in 1985, when wildfires destroyed over 600 houses in Florida.

California's most destructive fire (prior to Oakland Hills in 1991) hit the Santa Barbara area in June 1990. The Painted Cave fire started during a period of extreme weather; temperatures reached 109 degrees and humidity was down to 9 percent. It burned 641 structures as it moved west out of the coastal hills across 4,900 acres. Other parts of southern California were experiencing similar "fire weather," and within three days of Painted Cave, three more large wildland fires raged through the region. In 1991, with almost 3,000 residences destroyed and 25 deaths, the Oakland Hills fire took over the dubious distinction as the most destructive.

DEFENSIBLE SPACE

In the East Bay hills, goats are used now to browse the fuels and reduce fire risks for people living at the edge of the wildland-urban interface. In the City of Los Angeles, an ordinance adopted in 1981 required 100 feet of clearance from brush land to a homeowner's structure. Some citizens grumbled about the "loss of habitat" and the maintenance work required, but a 1988 report could point to no dwellings lost to brush fires in the Mountain Fire District in the prior seven years.[24]

The Los Angeles County Fire Department increasingly used prescribed fire to reduce fuels and also to train its seasonal crews for the wildfire season. "Rx Burn Manager" Captain Joseph Lindaman headed that program in 2000, when they introduced a one-of-a-kind "brush crusher."[25] Lowering and raising a ten-ton, fourteen-foot-long roller (from a bulldozer at the ridgetop), the tool crushed shrubs without destroying their root crowns. The crushed material would die and dry out. Then it would burn at times of the year when the rest of the surrounding vegetation was too green and moist to carry fire. Most stumps of species that would naturally resprout after fires survived. The crusher could be used on hills too steep for other equipment, produced less soil disturbance than bulldozers, and was expected to increase the number of days for safe burning. Safety was a major emphasis for Lindaman's program in that heavily populated county. Though they would prepare sites so that a small crew could handle the burns, then they would add, for every burn, 150 fire fighters to boost the safety factor.

A few months after the "brush crusher" was unveiled to the media, I asked for Captain Lindaman's opinion about a pamphlet sent to some southern California homeowners titled "Evacuating Topanga: Risks, Choices and Responsibilities." The booklet received news media attention for recommending that able-bodied adults *not* evacuate during wildfires, but instead, retreat inside their homes during the minutes that the area's windblown fires passed over, then go out and extinguish embers on roof or residual fire by walls. "Australia's doing that," Lindaman said, referring to a program called "Prepare, Stay, And Survive." "They quit taking people out; evacuating. Because, when do you lose people? When they're evacuating. They don't require them to stay, they tell them we'd like you to stay and if you do, here's what you do. You *have* to have proper clearance," he emphasized. "You *have* to prepare as if you don't have a fire department because, in those situations, you don't."[26]

Populations kept growing and sprawling into wildlands across the nation—not just within California and Florida, where the population pressure was particularly intense. Bad fire seasons not only increased interest in fuels management, but also increased calls for individual responsibility. *Must* the government or insurance agencies cover the costs of "natural" disasters? Should zoning laws restrict new subdivisions on fire-prone landscapes? Should society finance new roads and infrastructure that *encouraged* people to settle on fire-adapted landscapes? Might humanity itself figure out, instead, how to become one of the "fire-adapted" species?

Programs to teach people how to prepare "defensible space" around their homes and property have become a major emphasis of fire prevention campaigns. "Living With Fire," a guide for homeowners, was published in 1998 for communities in the Great Basin.[27] Provided as a special newspaper insert from Nevada to New Mexico, the publication showed homeowners how to determine the size of defensible space dictated by their local vegetation and terrain and advised them to remove dead material, thin shrubs and trees, clear away "ladder fuels" that might feed fire up into trees, keep the area within thirty feet of their houses "lean, clean, and green," and maintain those conditions over time.

"*It Can't Happen to My Home. Are You Sure*?" a U.S. Forest Service pamphlet asked.[28] Part of a national *Firewise* campaign, the pamphlet provided similar recommendations for treating the surrounding landscape. Houses themselves could resist fire if wood shingle roofs and siding were treated with fire retardants (better yet as roof materials were metal or composition shingles, with higher fire resistance without special treatment). Attic vents should be screened to keep out blowing embers and open decks and under-flooring should be enclosed. Firewood should be stackcd away from the house. Homeowners who kept their own hoses and nozzles might save their own property when fire fighters were busy elsewhere.

"If you build houses out of forest materials, they're going to burn just like forests," Stephen Pyne said during a NPR radio program. "The urban wildland problem is in some ways a dumb problem to have because it has a technical solution. We can reduce these large damaging fires . . . with zoning and building codes, simple housekeeping and simply building up some fire protection system. . . . And that's a different problem than trying to deal with fires in the wilderness area or trying to deal with very extensive wildlands that have been profoundly upset by changes in fires' patterns."[29]

In 1993, fourteen southern California wildfires destroyed 537 homes. The California fire marshall at the time was Ron Coleman. He toured the destruction caused by the Laguna Beach fire and gave traumatized victims advice about changes that should be made as rebuilding began. One woman whose house burned down told him: "Don't give me any of that fire prevention crap. I want it to grow back the way it was before." Coleman related that story in his keynote speech at the 2000 Fire Ecology Conference in San Diego, and also said "It has been my exposure in the last 40 years in the fire service that we hardly ever win at anything unless we have a burnt sacrifice to show somebody. You tell people what needs to be done; they tell you it's too expensive. Then there's a crisis and they suddenly start throwing money at it."[30]

Each year's new fire victims (unless they were stubbornly oblivious) had to admit that repeatedly they had been told "what needs to be done." Perhaps the most persistent voice, for over forty years, had been that of Harold Biswell. "Doc" Biswell died, January 7, 1992, two months after his comments about the Oakland/Berkeley Hills fire were televised. In February 1994, 350 people gathered in Walnut Creek, California, at The Biswell Symposium entitled "Fire Issues and Solutions in Urban Interface and Wildland Ecosystems."

The conference honored Biswell's interests in both wildland and urban fire issues and brought together professionals from both areas in a common forum. One point emphasized in James Agee's memorial dedication to Biswell was the importance "Doc" had placed on publication. Biswell authored 162 scientific and popular articles and bulletins on his research. He was the first editor of the *Journal of Range Management* in 1948, associate editor of the *Journal of Forestry* from 1950 to 1955, and associate editor of *Ecology* from 1956 to 1958. Biswell's book, *Prescribed Burning in California Wildlands Vegetation Management*, was published by the University of California Press in 1989. It is "a classic integration of science and interpretation," Agee said. "Harold took a complex problem and presented a complex answer, but in a way that most people could understand."[31]

Harold Weaver, Biswell's collaborator and friend, had died in 1983. "Not many folks remember him and that is probably a shame," Preston Guthrie told me. Guthrie was in charge of the Bureau of Indian Affairs resource planning branch in 2000. He never met Weaver, but the agency memory is of

> one of the more interesting and colorful characters in Indian Forest history. Some might say Weaver was simply a pyromaniac that was fortunate enough to find a relatively legal outlet for his fetish with fire in the rather liberal domain of Indian Forestry in the mid-decades of the 20th century. It is hard to dispute the man had a streak of brilliance running through his eccentricity.
>
> [Weaver] was very much aware of the impact man had on the environment and seemingly prophetic in his thoughts that this would be important in the future. Fortunately some of his work remains. Unfortunately many egocentrics of today possessing only limited knowledge of the past, do not realize they are reinventing the wheel, relative to fire use.[32]

We can only speculate about how the two Harolds would have reacted to events in the record-breaking fire season of 2000.

NOTES

HB—Harold Biswell papers, held at the Bancroft Library, University of California, Berkeley (BANC MSS 2002/67 c).

DPR—California Department of Parks & Recreation, Resource Protection Division files.

1. Bill Crosby, "Our Wild Fire: History Shows that Nearly all of California Is Designed to Burn," *Sunset* (June 1992), 65. Special printing for USDA, Forest Service, Pacific Southwest Region.
2. Margaret Sullivan, *Firestorm! The Story of the 1991 East Bay Fire in Berkeley* (Berkeley, CA: City of Berkeley, 1993), 9, 23, 33.

3. Ibid.
4. Firewise.org. 2000. "The Oakland-Berkeley Hills Fire." Online: http://www.firewise.org/pubs/theOaklandBerkeleyHillsFire/abstract.html
5. Biswell to unspecified recipients, February 2, 1971, HB.
6. Sue Soennichsen, "Must Hills Be Logged Off?" *The Montclarion*, February 7, 1973.
7. Norm Hannon, "Big Trees 'Freeze' Poses Problems," *Oakland Tribune*, February 11, 1973.
8. Richard Minnich, *The Biogeography of Fire in the San Bernardino Mountains of California* (Berkeley: University of California Press, 1988), 45.
9. William Mulholland, *Los Angeles Times*. September 12, 1908.
10. Crosby, "Our Wild Fire," 65.
11. "California Aflame! September 22–October 4, 1970," California Division of Forestry, Sacramento, November 1971, p. 67.
12. Leslie Connolly, "Chaparral Mismanagement," *Sierra Club Bulletin* 59, no. 5 (May 1974): 10, 11, 30.
13. Harold Biswell, "Prescribed Fire as a Management Tool," presented at the Symposium on Environmental Consequences of Fire and Fuel Management in Mediterranean Ecosystems, Palo Alto, California, August 1–5, 1977, p. 6 of Biswell's speech notes, HB.
14. Biswell to Clinton B. Phillips, September 28, 1977, DPR.
15. Douglas R. Leisz and Carl C. Wilson, "To Burn or Not to Burn: Fire and Chaparral Management in Southern California," *Journal of Forestry* February 1980): 94–95.
16. Serena C. Hunter, "Chaparral Fires: Are They Inevitable?" *American Forests* (September 1981): 25–27, 44, 45.
17. California Department of Forestry and Fire Protection, Vegetation Management Program, July 2000. Online: http://www.fire.ca.gov/ResourceManagement/VegetationManagement.asp
18. Joe Rawitzer, Advanced Fire Technology (a private firm), November 27, 2000, personal communication.
19. James D. Brenner, Florida Division of Forestry, May 21, 2001, personal communication. Also see Florida statute 590.125(3)(c) and Florida Administrative Code 5I-2 (2). Online: http://flame.fl-dof.com/Env/law.html
20. Scott E. Franklin "Prescribed Burning at the Urban Interface in Los Angeles County," Chaparral Resource Management, Newsletter No. 3, Summer 1984, p. 4.
21. S. E. Franklin, "LACFD Weeds Out Dangerous Fuels with Vegetation Management Program, *American Fire Journal* (January 1988), 24–25.
22. Richard A. Minnich, "Fire Mosaics in Southern California and Northern Baja California," *Science* 219 (March 18, 1983): 1,287–1,294. The copy with Minnich's personal note is in Harold Biswell's papers (HB).
23. John T. Koehler, "The Use of Prescribed Burning as a Wildfire Pre-

vention Tool," Florida Division of Forestry, Orlando. Online: http://flame.fl-dof.com/Env/koehler.html#lit. Accessed May 20, 2001.

24. J. O. Haworth, Contemporary Fire Prevention at the Wildland/Urban Interface, *The Fireman's Grapevine* (October 1988): 14–17.

25. Los Angeles County Fire Department, "Brush Crusher," press conference, news footage videotape, Pacoima, California; "Prescribe Burn Program with Capt. Joe Lindaman"; and Los Angeles County Fire Department, videotape, June 2000.

26. Joseph Lindaman, Los Angeles County Fire Department, interviewed November 28, 2000, San Diego.

27. University of Nevada, Reno and Sierra Front Wildlife Cooperators, *Living with Fire*, May 1998. Great Basin version funded by Great Basin Fire Prevention Organization in *Inyo Register*, Bishop, CA, July 1, 2000.

28. U.S. Forest Service, *It Can't Happen to My Home. Are You Sure?* (Washington, D.C.: Government Printing Office, 1999), 1999–773-302/24125.

29. Stephen Pyne on "Talk of the Nation," National Public Radio, hosted by Juan Williams, August 14, 2000.

30. Ron Coleman, *Fire Ecology Conference 2000*, San Diego, author's record of speech, November 28, 2000.

31. James K. Agee, "Memorial Dedication to Dr. Harold H. Biswell," in *The Biswell Symposium: Fire Issues and Solutions in Urban Interface and Wildland Ecosystems, February 15–17, 1994*, Rep PSW-GTR-158. USDA Forest Service Gen. Tech. Rep PSW-GTR-158, Walnut Creek, California, 1995, p. 3.

32. Preston Guthrie, personal communication, December 21, 2000.

12. Escape!

We have caught the bear by the tail—can we let it go?
Stewart Edward White, 1920[1]

The unnatural intensity of twenty-first century wildfires brought the threat of permanent environmental destruction and an unacceptable casualty rate to the warriors America sent into battle against wildfires. After thirty-four fire fighters were killed in 1994, an interagency review of federal wildland policy produced a unified national policy for the first time, recognizing that fire was a critical natural process. The National Fire Plan[2] required fire management plans for every area with burnable vegetation on federal land. It emphasized safety and indirect attack strategies to decrease fire fighter deaths. Flexibility was permitted at last, so a full-range of responses could be considered for every wildfire, from basic monitoring to low-impact confinement up to full-scale suppression efforts.

Agencies more than doubled the acreage treated with prescribed fire between 1995 and 1999, yet the 2.2 million acres of federal land burned in 1999 remained well below what was needed to rapidly restore balance on 200 million acres of federal wildlands that were ecologically adapted to frequent fires. When the wildfire season of the year 2000 arrived, it severely tested the commitment to national fire management. An escaped prescribed burn led to a wildfire that destroyed hundreds of homes in Los Alamos, New Mexico.

THE WAKE-UP CALLS

A special report in September 1999 in *Forest* magazine focused on the possibility of a forest fire "inferno" burning into the Los Alamos National Laboratory (LANL). "Two major wildfires have broken out over the past three years . . . in a spot just a mile or two from about the last place in the world anyone in their right mind would want to see a forest fire: Los Alamos National Laboratory [LANL], birthplace of the atomic bomb and home to a witch's brew of radioactive materials and toxic chemicals."[3] The fires were the Dome and Oso Fires of 1996 and 1998. The La Mesa Fire had also burned to within a few miles of the community and laboratory in 1977. It was a matter of weather and luck that none of those wildfires burned into the LANL or the town of Los Alamos, where 12,000 people lived.

After the La Mesa fire scientific symposiums were held in 1981 and 1994. The Santa Fe National Forest, NPS, LANL, the Department of Energy (DOE), and Los Alamos County collaborated on plans to reduce wildfire hazards. Fire breaks, thinning, and some burning had been done around the boundaries of the laboratory and the NPS began burns within Bandelier National Monument. Bandelier, established in 1916, was one of the first national park units; its boundary was just southwest of Los Alamos.

Craig Allen, a fire ecologist with the U.S. Geological Survey's Jemez Mountains Field Station at Bandelier National Monument, considered the town and laboratory to be "on board," but saw a potential for massive crown fires. "To get the town's attention, we expanded fire scar research data sites into the heart of the town, to canyons that run right through the middle of the community. We found that the last fire was in 1881."[4]

Los Alamos sits atop a series of mesas surrounded by ponderosa pine forest. In some areas there were 2,000 ponderosa pines per acre where just 25 to 80 were found back when frequent low-intensity fires burned. West of the laboratory and town, on Santa Fe National forest land, the special report in *Forest* magazine described "a scenic but ominous backdrop to the lab and the community of Los Alamos"—a dense forest rising to the peaks of the surrounding mountains.

Fire history studies by Craig Allen and Tom Swetnam, a dendrochronologist from the University of Arizona Laboratory for Tree Ring Research, documented fires about every ten years in the eighteenth and nineteenth centuries.[5] Though the Jemez mountain range received the nation's second highest rate of lightning strikes, fire occurrences declined in the late 1800s when settlers began grazing large flocks of sheep. They cropped away the grass that had carried "flashy" fires along the ground, but seldom led to crown fire in the trees. Then, in the twentieth century, fire suppression policies took hold. Thick trees closed in on the former savannah-mix of grass and trees. High-severity crown fires became the norm.

On April 26, 2000, two weeks before Bandelier National Monument would ignite a prescribed fire on Cerro Grande mountain, a public meeting, titled "Wildfire 2000: Los Alamos at Risk," was cohosted by the Los Alamos National Laboratory and the Interagency Wildfire Management Team (IWMT). "It's not a matter of if but when wildfire will again threaten the Lab, Los Alamos and surrounding areas," the IWMT chair, Diana Webb, said. "We can't stress this enough."[6] Despite thinning, areas of the laboratory and town remained at high risk. The forecast was for an exceptionally dry season.

Thursday Morning, May 4

Burn boss Mike Powell notified the Santa Fe Zone Dispatch of the intent to ignite a prescribed fire that evening, near the summit of Cerro Grande mountain, a 10,190-foot peak within Bandelier National Monument. The dispatcher told him to be careful because conditions were dry.

The burn was meant to keep fire up on about 300 acres near the summit of the mountain. With that crest as a treated buffer, the rest of the 1,000-acre project area farther down hill could be burned later, in two separate phases. On May 1, three days before the burn was ignited, there had been a dusting of snow on the peak. Though the region was in its third year of drought, live and dead fuel moistures being monitored showed that fuels on the mountain were in prescription. Bandelier officials decided "that the weather and moisture conditions in the area of the burn . . . were more favorable than the publicly available information suggested."[7]

THE CERRO GRANDE FIRE

At 7 PM, from the summit of Cerro Grande, the "holding boss" (in charge of crews monitoring the fire once it was lit) called the National Weather Service on a cell phone to reconfirm a "spot weather forecast" that had been faxed to the park earlier in the day. Routine fire weather forecasts were issued twice a day, but burn managers could also receive the site-specific "spot" forecasts. Extended wind forecasts, for the next three to five days, were *not* part of the forecasts available from the Weather Service. A few minutes after that phone call a "test fire" was ignited. Flames behaved as expected, so about 8 PM the twenty-person crew began to create a blackline along the northeast edge of the plot. Drips of flame from the diesel-gasoline mix in drip torches lighted a strip of fire that was only allowed to burn downhill. Once a blackline was begun, strip fires were ignited a short distance farther downhill to burn up toward the expanding blackline. They were burning in grass at first. Soon they began to let fire back freely down the slope, contained by the blackline uphill to the east.

Half way along the eastern edge of the area, fuels changed from grass to a mix of leaves, needles, branches, twigs, and logs. A fire line had been previously constructed along that boundary of the parcel, down the mountain to a highway, State Route 4. They did not plan to bring fire down as far as the highway in this first phase of the project. Ignition on the east line was stopped about two hundred yards above the fuel transition.

About 11 PM, with ignition complete along the northeast edge, some fire was noticed outside the blackline. That 30-by-30-foot escape was soon suppressed. They began igniting the other, northwestern edge of the parcel at about 11:15 PM. As that line wrapped around from the summit to head down toward the highway, fire would burn down toward the other flames backing from the east.

May 5

During the night the Black Mesa (BIA) crew began experiencing extreme fatigue from the high altitude. Though they had been expected to work all night, by 2:30 in the morning, for their safety, they were sent down the mountain. Six of the park crew stayed to monitor the burn. The burn boss also went down to the park office to order a replacement crew and a helicopter. He called Zone Dispatch at 3 AM, but was told he would have to call back in the morning. "Contingency resources" that had been identified in the burn plan as available within one-and-a-half to three hours, he now learned, could not be immediately assigned. Authorization had to come from a supervisor who would not be on duty for several hours.

When Powell reached the dispatch supervisor at about 7:30 in the morning, there was further confusion about assigning the hand-crew and helicopter. The dispatcher's understanding was that contingency resources could not be sent to a prescribed burn unless it had escaped and been declared a "wildfire." They worked out a way to assign the resources[8] and, finally, at 11:30 AM, a helicopter dropped off two fire fighters. A thirteen-man fire crew reached the mountain at 12:30 PM, more than nine hours after the first call had been made.

In the meantime, on the northeast side of the parcel there was a "slopover" that the small holding crew had trouble containing. They needed water drops and support. With those requests, the burn was officially declared a "wildfire" and an air tanker was dispatched. It arrived about 12:55 PM to make water drops on the northeast side of what now was called the "Cerro Grande Fire." Efforts from that point were all aimed at suppressing the fire.

Paul Gleason, a wildfire management specialist for the NPS, took over as Incident Commander (I.C.). He happened to be at Bandelier National Monument to look at the lower end of Frijoles Canyon, farther down the mountain, where a more complex burn in denser fuels was planned as a later phase of the overall project. One option would have been to confront the fire along the south edge of the upper parcel, confining it within the 300 acres planned for that burn. Gleason consulted with his fire management team, then made what would be a pivotal decision. He felt that working across the face of the mountain was dangerous; it would require line to be "underslung"—scraped and cut beneath overhead trees, some with dead limbs. He made a decision to "go indirect," instead.

"That's a point of controversy and believe me I've gone over that decision thousands of times since," Gleason later said. "Because of the problems the hotshot crew was having on the underslung line, because the fire at this time was starting to get into this old growth, decadent tree and snag area, and because ten years earlier in a North Cascades park, I put one of my squad bosses in the hospital because of trying to do underslung line—a tree fell, a limb went cartwheeling through the air, and hit him in the chest—I gave to the park superintendent, as my preferred alternative, to go indirect, following the lines that had previously been developed, down to State Route 4. . . . Critical decision, I revisit every day."[9]

Crews began improving the existing lines and blacklining along them with drip torches, down toward the highway.

May 6

Through the night, blacklining continued along the east and west sides to stay ahead of the main fire as it continued backing downhill. At 7:30 AM another spot fire beyond the burn boundary to the east was contained. Saturday brought more fire fighters and several retardant drops were made. That day the NPS fire managers met with County, LANL, and Forest Service personnel. On Saturday night, as they worked to complete the burning down to Route 4 on the west side, they encountered wet meadows in the lower portion of the west side that were hard to ignite.

May 7–11

On Sunday, at around 2:30 in the morning, the fireline was tied to its "anchor point" at Route 4. The next goal was to bring the blacklining fire across, east to west, but winds were blowing the wrong direction—down slope. Across the highway to the south, was Frijoles Canyon, choked with fuel. So they waited, planning to finish that final blacklining in the evening. The general feeling among those managing the fire was that they were on top of it, and about had things "sewn up." At about 10 AM a helicopter attempted to widen out the fire line along the west line by igniting fire to burnout more fuels.

Two hours later, at noon, winds suddenly increased from the west. Gusts were measured up to 50 miles per hour at the Nuclear Laboratory. The wind fanned embers and began spreading fire rapidly from the west to the east, paralleling Route 4. A large spot fire was ignited *south* of the highway in the heavy fuels of Frijoles Canyon. That new fire burned too intensely for crews to attack it at first. It became a crown fire that sent embers into the wind that started more spot fires and crowning to the east. The Frijoles Canyon spot fire itself was contained about 5 PM, but fire ran across the east face of the mountain, with 100-to 150-foot flame-lengths.

Headlines in the *Los Alamos Monitor* chart the events that followed, as new wind "events" kept frustrating a growing army of fire fighters:

"Prescribed burn escapes boundaries" (Sunday, May 7)

"Wildfire! Worst fears become reality for Los Alamos" (Tuesday, May 9)

"Los Alamos Evacuated!" (Wednesday, May 10)

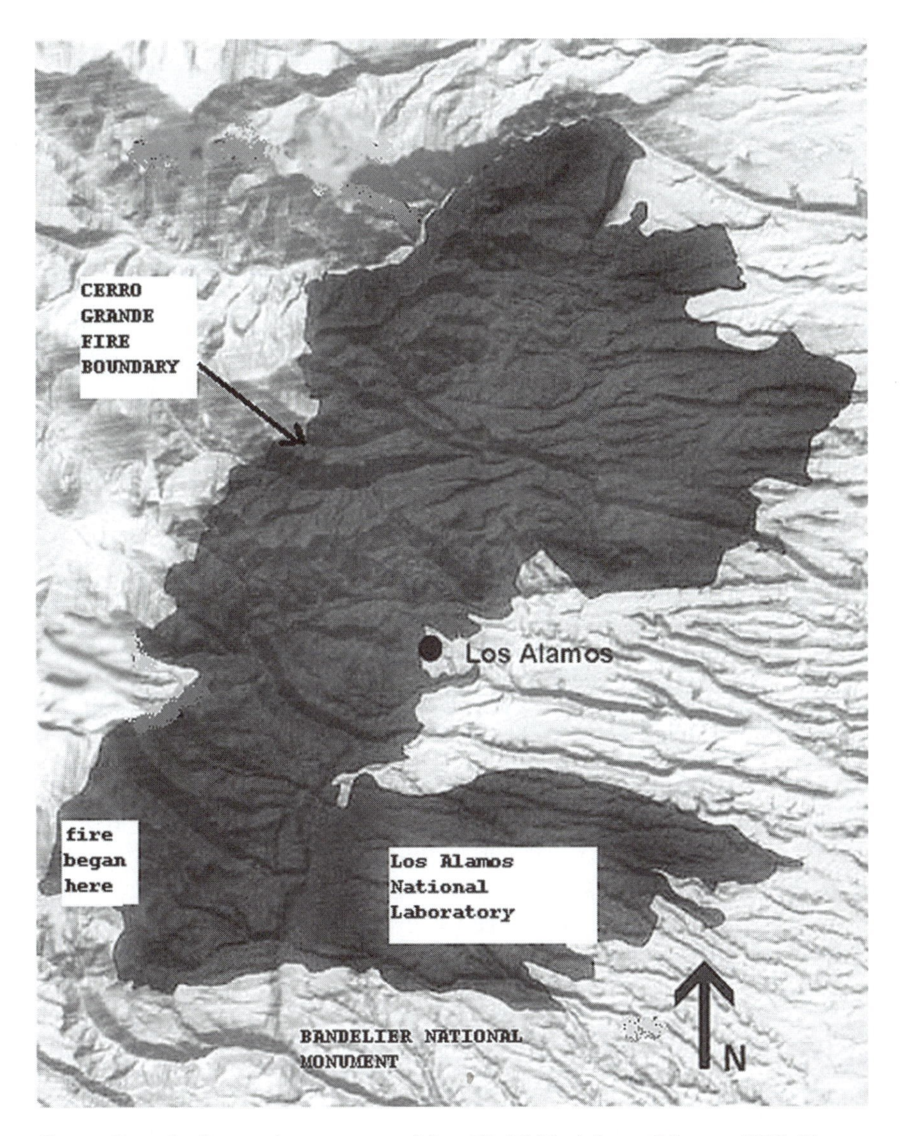

Cerro Grande fire perimeter map, May 17, 2000. Adapted from a U.S. Forest Service map.

Between about 5 PM on Wednesday night, and the early morning hours of Thursday, the fire burned 20,000 acres. Hundreds of homes were destroyed in Los Alamos, as the main body of the fire moved on north toward the San Ildefonso and Santa Clara Indian pueblos.

The Cerro Grande Fire was finally declared 100 percent contained on June 7. Eighteen thousand residents had been evacuated from Los

Homes burned on the edge of Los Alamos by the Cerro Grande fire, May 2001. Courtesy of Los Alamos National Laboratory.

Alamos and White Rock; 239 homes burned down, some that were muliple-dwelling units. Over 400 families were displaced. Thirty-nine Nuclear Laboratory structures (office trailers and sheds) were destroyed or damaged, but there were no radiation escapes. Damage costs were estimated at more than $1 billion. The only good news was that no one had been killed and no fire fighter was seriously injured.

"The dejection in Mike Powell's voice was palpable," a Santa Fe newspaper reporter wrote, on May 11. "Powell was the burn boss—the guy who co-wrote the burn plan for a fire that was supposed to be controlled but has become an inferno . . . 'I feel extremely bad about it, and it's not just me—it's this whole park', Powell said Wednesday of the staff at Bandelier National Monument. 'There's a lot of things happening. It's pretty emotional. People are teary', said Powell, his voice sometimes halting. Bandelier Superintendent Roy Weaver said, 'Based on what we knew at the time and what we believed had to be done, we think we did what was right.' Weaver said he knows a lot of people are angry, worried, frightened, frustrated and inconvenienced. 'They have a right to be frustrated and angry,' he said."[10]

AFTERMATH

"Fire Victims Return to the Scene of the Disaster" was the *Los Alamos Monitor* headline on May 16. Don and Elain Morris, their daughter, Dawn, her husband, Louis Jalbert, and their young daughters lost their homes in side-by-side units. After they had evacuated, the Jalberts had seen televised images of their home burning. They were allowed back into the area on May 12. Louis Jalbert, a hazardous waste sampler at the Nuclear Laboratory, told a reporter that he had felt vulnerable and, in a way, was not surprised by the fire. "Jalbert said he thinks mistakes were made in Bandelier National Monument's controlled burn, but he still believes such fires are a good policy. He said he's not angry and feels that the fire was an act of nature in a tinderbox."[11]

Not all the victims were so philosophical. A Los Alamos resident's letter to newspapers suggested that "the officials who made the decision to 'put *our* lives, *our* town and *our* mountains in extreme danger" be removed from office and 'be forced to serve mandatory jail time with all the common criminals.' "[12]

Bandelier Superintendent Roy Weaver took responsibility for the fire and retired a year earlier than he had planned, on July 2. National Park Service employees were not allowed to discuss the fire or react to information in the media while investigations were underway. On July 13, as a private citizen, Weaver's apology to the community was published. It read, in part:

> Dear Neighbors . . . Now that I am retired from the NPS, I can express myself as a private citizen. I hope it is not too late.
>
> A tragedy did occur that hurt so many of you in so many ways. Homes can be replaced but so many keepsakes and personal treasures—pieces of your lives—that were lost, cannot be replaced and are now only memories. I can identify personally with that loss.
>
> I truly regret the pain and loss, the anguish you have had to suffer as a result of the Cerro Grande fire and wish every day that with some sort of magic, I can make it all go away. . . . Your kindness to each other has been inspirational, and your kindness to my wife and myself has been overwhelming. We have received messages of encouragement and support from several hundred of you. Messages that, while still cognizant of the tragedy that struck so many of you, want us to know that we are still loved and respected neighbors.[13]

As the nation sympathized with devastated homeowners, the need to reexamine policies and procedures was obvious, but the conse-

"Credentials," an Oliphant cartoon. © Universal Press Syndicate. Reprinted with permission. All rights reserved.

quences of a century of war on wildfire were also part of the context as Dr. Stephen Pyne wrote: "If the right fire is the right thing for the land, then we have to find a way to do it, and not pretend that bad fires are good. We can't allow feral fire to roam through town like ravenous grizzlies."[14]

QUESTIONS AND INVESTIGATIONS

Secretary of the Interior Bruce Babbitt formed an interagency Fire Investigation Team on May 11 to examine the escaped burn. The team was given just one week to prepare their report. Secretary Babbitt and Secretary of Agriculture Dan Glickman also suspended all federal prescribed burning west of the one hundredth meridian (a line just east of the Rocky Mountains) for thirty days. The Fire Investigation Team concluded "that federal personnel failed to properly plan and implement the Upper Frijoles Prescribed Fire, which became known as the Cerro Grande Prescribed Fire. Throughout the planning and implementation, critical mistakes were made" regarding the park service's complexity analysis process, review of the plan, evaluation of conditions outside the fire boundary, provision for adequate contingency resources, and safety policies. "The investigation team believes that the Federal Wildland Fire Policy is sound; however, the success of the policy depends upon strict adherence to the implementation actions throughout every agency and at every level for it to be effective."[15]

Secretary Babbitt had a four-person independent board review the investigation report. Robert Stanton, director of the NPS presented their findings to the Senate on July 27. The board supported the findings of the investigation report and added, "Even with a continuing commitment to the use of prescribed fire, it will take decades to fully reverse hazardous fuel accumulations and unnatural ecosystem changes that have developed in the last one hundred years. Many areas may require more than one treatment either through prescribed fire or by mechanical means, or both, before fuels and ecosystem components stabilize. Many areas will have to be treated periodically, forever."[16]

Another investigation was completed by the General Accounting Office (GAO). On July 20, their report, "Lessons Learned from the Cerro Grande (Los Alamos) Fire," was presented to the Senate Committee on Energy and Natural Resources. "For prescribed fires to continue to enjoy public acceptance and remain a viable tool, the lessons of this fire cannot be lost."[17] The GAO report called for prescribed burn plans to be reviewed by independent, knowledgeable individuals, clarification on the process for acquiring contingency resources, and better coordination between agencies. "The overall complexity of the burn and the resources needed to keep it under control were underestimated. The effect of having insufficient resources snowballed until the fire was out of control." The investigators recognized that "This kind of analysis has the benefit of hindsight. We did not have the burden of making urgent, on-the-spot decisions in the midst of trying to manage an ongoing fire. Accordingly, we are not here to assign blame but to help improve the way federal land management agencies manage future prescribed burns."

At a national fire ecology conference in November 2000, a panel of speakers discussed the Cerro Grande fire. Researcher Craig Allen, whose data helped in the design of the fire plan, said: "The management at Bandelier wasn't just playing around on that mountain. They felt that there was a sense of urgency to try to reduce those fuels.

Management at Bandelier was extraordinarily conscious of the risks posed by ecological changes in the landscape. NPS knew very well the powder keg nature of the fuels, prevailing winds, etc. The park knew that [its vegetation] was the fuse for that powder keg. The goal and sense of urgency was to reduce the fuels."[18]

Paul Gleason, the incident commander in the first days when the burn was declared a wildfire, said:

> I'm assuming that a lot of people in this room are carrying the torch, and you're making some critical decisions and been put in some tight spots, and the reason you're sitting here, instead of going over to the San Diego zoo, is you're scared to death that this could happen to you. The intent that I have is, one, to get the record straight. When the initial investigation came out they had an extremely short period of time. To me, more important, instead of laying blame, is to find out what happened and how we can keep it from happening in the future, because we've got a lot of fuels. We're done burning the grasslands and the easy areas, now we got some tough stuff to deal with.
>
> People say when it was declared a wildland fire, that it was being managed for resource benefit. And it wasn't. It was firefighter safety is what caused me to go to the park superintendent and say, "These are the two choices we have, direct or indirect." I personally liked the indirect approach because we wouldn't be putting people underneath fire weakened snags with the potential of getting hurt.
>
> To me, to see a report that said there were no escape routes bothered me more than anything else. If they said I was a pyromaniac, anything else, that'd be ok, but that I wasn't conscious of firefighter safety, that bothered me.[19]

The GAO report had noted that, during the fire, none of the on-site fire-fighting officials expressed concern about the burnout technique that introduced fire along the west side of the burn area. "However, after the fire, the on site Park Service official who was in charge recognized that this was a tactical error and said that if he had better information on the wind for May 7, the day the fire began to move towards Los Alamos, he would not have introduced fire into the western portion of the burn area. As things unfolded, the introduction of fire in the western portion of the burn area led directly to the fire's getting out of control on May 7."[20]

Paul Gleason was the official referred to in that paragraph. He wanted to clarify that decisions were made with the best information available. "You know, I don't feel it was an *error*," he told me. "I had gotten information about the wind speed . . . but the information I had wasn't telling me that we were going to get 50 mph winds that afternoon."

There is irony in the fact that the fire that was spread by the wind originated, not with the prescribed burn, but from fire introduced as a suppression tactic. Forest Service fire behavior analyst John Robertson explained, in an appendix to the investigation report, the chain of incidents that led to that choice and its tragic results. It began on the first night, with the crew that was unable to stay on the burn:

> It is probable that if contingency resources were at the burn site on May 5 these resources would have been able to contain the slop-over without the need to convert the prescribed fire to a wildfire. Instead, the prescribed fire would have progressed to the timber fuels where it is probable that ignition of the lines would have slowed or stopped completely. . . . [Because the timber fuels were hard to ignite] it is very unlikely that fire would have spread down the west line in to the flats (along Road 4). . . . Even if fire had managed to work its way along the west line it would have done so slowly and been easily contained. . . . there would have been no fire approaching the road and no need to burnout along Road 4. . . . fuels would not have been preheated and dried out and no ignition source would have existed to initiate the crown fire that resulted in the spotting outside the project area. . . . The strong winds do not appear to have created active fire spread in the grass fuels or timber fuels that had been burned the night and early morning of May 5 (upper third of the project area). There is no indication that the source of the escape fire came from this area of the burn.[21]

Paul Gleason's career began on the Angeles National Forest when he was eighteen years old and became a seasonal fire fighter. He had moved from Iowa to southern California when he was in the fifth grade. "In the sixth and seventh grade I'd ride my bike up into Dalton Canyon and the firefighters up there, on Sunday afternoon, they'd let me sit on their fire trucks; I knew I wanted to be a firefighter." He worked on Mt. Hood National Forest in the early 1970s, when more prescribed fire began to be done, working "hotshot crews" during the wildfire season, but burning clearcut units in the spring and fall. He went to graduate school in 1991, studying ecology and environmental ethics. "I took a lot of philosophy," he told me, "because it was important in the Front Range to bring this message out to people on how important fire was as a process." About two years before the Cerro Grande fire, he began working in Denver for the Rocky Mountain region of the NPS as a wildland fire management specialist, doing support services for individual parks.

When the "Type I" incident command team took over on the Cerro Grande fire, after it escaped and raced east toward Los Alamos, Gleason served as a NPS representative. It was, he said, "real hard to go to Santa Clara pueblo and sit in an auditorium of the Santa Clara tribe and stand up and say, 'My name's Paul Gleason. I work for the National Park Service and we're the ones that lit this fire, and I'm sorry from the bottom of my heart about what has happened. I was the Type 3 I.C. [incident commander] when it blew off the face'. That was really hard to do, but I sleep better at night now, rather than try to hide or deny it. So that's one thing, is accepting responsibility for what's happening." Knowing that no lives were lost helps him deal with the aftermath of the tragic loss of homes and property. "I get through it, because I really do believe in firefighter safety."

The primary "powderkeg" is still in place at Bandelier National Monument. The fuel level in Lower Frijoles Canyon, the complex area that brought Paul Gleason to Bandelier, was called "explosive" by several conference panelists. The situation there was not corrected by the Cerro Grande fire. I asked Paul Gleason what could be said to people in Los Alamos about the *next* burn that remains to be done.

"In real critical areas like that we're going to need to do mechanical treatments to reduce fuels with chain saws. Now, I'm one real strong advocate that there's a lot going on when you burn fuels, because you're releasing and recycling nutrients. That isn't achieved if you just pull the log out of there. And, really get the community to buy off on it, for the Park Service to not accept lone responsibility for their project. It's a community project, maybe on park service land, but a community project. And the stakeholders need to say, 'Yes, we as citizens of the communities that lie downwind from that firespread under the worst case scenario are willing to accept the risk.' " Though he was about to retire, Gleason said he would volunteer to help with that next burn if recommended changes were implemented, despite his experience that year—or perhaps because of it. "[Lower Frijoles Canyon] still exists as a highly explosive powder keg, and needs treatment or we're going to see this kind of situation again."

THE WAKE-UP CALLS REVISITED

Six months after the fire, black tree skeletons showed where intense crown fire had moved across the hillsides. So it was a surprise to see how many trees along the boundary between the forest and the town in Los Alamos were still alive, though with scorch marks climbing up their trunks and lower branches. Forest Service fire investigator Jack Cohen examined the area on May 14, 2000, and concluded that much of the fire burned "within several hundred yards or more of the Los Alamos residential area . . . as a surface fire—an underburn. . . . the tree canopy was variably scorched but not consumed. My examination suggests that the high ignitability of Los Alamos was principally due to the abundance and ubiquity of pine needles, dead leaves, cured vegetation, flammable shrubs, wood piles, etc. adjacent to, touching and/or covering the homes. . . . the high ignitability of most of the residential area allowed numerous simultaneous house fires that quickly overwhelmed the suppression forces."[22]

Tom Swetnam, director of the University of Arizona tree ring laboratory made another point: "There is plenty of evidence that these forests were going to burn like this sooner or later, whether it was started by lightning, careless smokers, or unattended campfires. There's no excuse for these kinds of mistakes being made," he immediately added, "but the great irony is that the Park Service and the Forest Service were trying to avoid these kinds of fires."[23]

As the anniversary of the Cerro Grande fire approached, a flurry of letters appeared in the *Los Alamos Monitor* anticipating the NPS Board of Inquiry report that would determine whether personnel action was warranted against participants in the event. Several letter writers claimed that persons from various agencies, including LANL, had pleaded with Bandelier officials to not conduct the burn during the weather conditions of early May 2000. Mat Johansen, of the LANL staff, answered that charge on May 11, 2001:

> I recall a May 4, 2000, meeting at which a Park Service representative described their intentions to start the burn at Cerro Grande that evening. The representative detailed, where, when, and their goals for the fire. Twenty-eight people heard this, including representatives from LANL, DOE, Los Alamos County, US Forest Service, nearby pueblos, and the New Mexico State Forestry Division. Not a single objection was raised. In fact nothing close to an objection was raised. Likewise at the Interagency Wildfire Management Team meeting held eight days prior to the fire, no objections [had been] raised.[24]

Former park superintendent Roy Weaver wrote the newspaper on April 3, 2001, saying that the recent letters deserved a reply. He repeated that no "pleas" had been made to the park. Weaver said that the letter writers deserved to be angry. He was looking forward to the long-awaited Board of Inquiry report, not because he expected it to absolve anyone of responsibility, but he hoped it would "show in detail that Bandelier had followed existing guidelines rather than casually and irresponsibly disregarding the guidelines for prescribed burning. There is a partial quote in the recent AP article," Weaver added, "taken out of context, in which I allegedly stated that I would approve this prescribed fire again if I had it to do all over again. That . . . implies that I am callous and have no regard for the pain suffered by the people of Los Alamos. That sound-bite left out the complete text in which I qualified that . . . by stating that if I knew then what I know now, I would not have approved the prescribed fire."

The NPS Board of Inquiry report, completed in February, was not released for several months so that it could be reviewed by President Bush's new administration. Finally made public on June 12, 2001, the report found that there had been "errors in judgment and planning," but stated, "the planning and implementation actions of the principals were not arbitrary, capricious or unreasonable in the light of the information that they had prior to the burn" and were in compliance with National Wildland fire policies.[25] As Weaver had predicted, the report confirmed that NPS guidelines had been followed by those who conducted the burn at Bandelier National Monument (while pointing to problems with the agency's complexity rating procedures, and a need to improve coordination between agencies both in planning and implementing burns).

SETBACK OR A NEW WILL TO MANAGE FIRE IN ECOSYSTEMS?

Wildfire remained in the headlines throughout the summer of 2000. A few weeks after the Cerro Grande fire was controlled, dozens of wildfires began to ignite across the West, particularly in the Rocky Mountain states. They ultimately burned over 8 million acres.

On August 14 National Public Radio's "Talk of the Nation" host Juan Williams opened his program saying: "More than 60 major wildfires are now burning at a rapid rate in the Western United States. Republican governors in the Western states blame the Clinton administration for not doing enough to thin forests and prevent the fires [and] for opposing timber cutting on federal land." Fire historian Stephen Pyne made the point, on that program, that "fire exclusion did not begin with the Forest Service or with fire suppression, it began really with overgrazing in the latter part of the 19th century . . . it was the removal of that grassy surface cover which began the process of removing fire. The real tragedy was not that we put fires out; it's that we quit lighting them . . . and now it's very difficult to just go back in."[26]

"Fox News Sunday" host Tony Snow interviewed Montana Governor Marc Racicot and Dan Glickman, the Department of Agriculture secretary, on August 27:

> *Snow*: The Western U.S. has become a giant tinderbox with blazes raging across parts of 10 states. Montana has taken the most punishment. At one point last week, 30 fires were *incinerating* more than 600,000 acres of land.
>
> *Glickman*: This is like "The Perfect Storm," the movie and the book. It's a convergence of all these things at one time. We have the hottest, driest weather in perhaps 50 years; we have thousands of lightning strikes an hour; we have 300 new fires every day in the West, largely because of lightning strikes. And I would point out we're putting out 98 percent of those fires. But the convergence of that hot, dry weather, . . . coupled with a policy of Congress and the administration, for nearly 100 years, of suppressing virtually every forest fire, has created an environment where the fuels that are available for burning are greater than they might have been 500, 600 years ago.
>
> *Racicot*: Let me make this point clear: We're not treating near the fuels that an agency within the Department of Agriculture has said needs to be treated every single year. We're talking about reducing fuels in forests, and that involves a number of different things. It involves prescribed burning in those frequent forest areas that have frequently had fire and need fire.[27]

That the escaped prescribed fire that burned into Los Alamos did not sound a death knell for efforts to restore fire on America's wildlands may best be explained by the "perfect storm" of wildfire holocausts that dominated the rest of that summer. Rather than shutting down programs for years, as happened after the Yellowstone fires of 1988, the NPS resumed some burning at the end of October. That fall the federal government authorized almost $2.9 billion dollars for proactive wildland fuel reduction and fire management. Prescribed fire was a key "tool" to be funded. The question became just how wisely the dollars would be spent.

Governors of western states designed the new program on wildfires

A crown-fire holocaust on August 6, 2000 forced two elk to shelter in the Bitterroot River, Montana. Photo by John McColean, BLM, Alaska Fire Service.

and forest ecosystem health; President Clinton signed it into law on October 11, 2000.[28] Nearly $30 billion would be needed over the following ten years to implement the National Fire Plan's goals of wildfire risk reduction and rehabilitation. A comprehensive strategy would make the states full partners in planning and implementation. On November 9 the Forest Service unveiled its new program: "Protecting People and Sustaining Resources in Fire-Adapted Ecosystems—A Cohesive Strategy."[29] Aggressive fuels treatment, with fire as a key tool, was to make ecosystems where fire had been excluded safer for firefighters and the public.

Considering the trauma experienced in New Mexico that season, the July 2000 editorial in that state's *Deming Headlight* may have been the clearest sign that there was no going back to total fire suppression and that a broader public understanding existed about fire: "The federal government should not end its controlled burn program. It should simply be more careful."[30]

A *Los Angeles Times* editorial pointed to a Forest Service warning that 39 million acres of their land needed urgent treatment. "That a blaze raged out of control in New Mexico," the editorial stated, "is not in itself a reason to end controlled burning." The *Times* considered that the NPS had a good record overall, with just 1 percent of 3,783 burns escaping in the prior two decades. "However, Los Alamos does show that the government does not have nearly enough trained staff to accomplish the fivefold increase in controlled burning that it had planned over the next five years." The editorial supported new funding coming from the Clinton administration. "The government, in the end, has little choice but to fight fire with fire."[31]

NOTES

1. Stewart Edward White, "Woodsmen, Spare Those Trees! Our Forests Are Threatened; a Plea for Protection," *Sunset, the Pacific Monthly* 44 (March 1920): 23–26, 108–117.
2. "Federal Wildland Fire Management Policy and Program Review," Report to the Secretaries of the Interior and of Agriculture by an Interagency Federal Wildland Fire Policy Review Working Group, December 18, 1995, National Interagency Fire Center, Boise, Idaho.
3. "Los Alamos Inferno: Los Alamos National Laboratory Has Long Played with Nuclear Fire. But Can It Handle a Forest Fire?" *Forest* (September/October 1999) http://www.forestmag.org/losalamosfire.htm
4. Craig Allen, "Panel Discussion: Cerro Grande Fire," The First National Congress on Fire Ecology, Prevention and Management, San Diego, California, December 1, 2000. Not yet published, personal record.
5. C. D. Allen, R. Touchan, and T. W. Swetnam, "Landscape-Scale Fire History Studies Support Fire Management Action at Bandelier," *Park Science* 15, no. 3 (1995): 18–19.
6. "Laboratory Co-hosts 'Wildfire 2000: Los Alamos at Risk' on April 26," Los Alamos Nuclear Laboratory press release, April 21, 2000. Online: www.lanl.gov/worldview/news/releases/archive/00–061.html
7. Barry T. Hill, *Fire Management, Lessons Learned from the Cerro Grande (Los Alamos) Fire*, GAO/T-RCED-00–257 (Washington, D.C.: General Accounting Office, 2000), 5.
8. The dispatcher assigned the resources to a wildfire that was being fought elsewhere, then shifted them to Bandelier.
9. Paul Gleason, "Panel Discussion: Cerro Grande Fire," The First National Congress on Fire Ecology, Prevention and Management, San Diego, California, December 1, 2000. Not yet published, personal record.

10. Shonda Novak, "Bandelier Staff Despondent Over Blaze," *The New Mexican*. Santa Fe, New Mexico, May 11, 2000, p. A-7.
11. Sarah Meyer, "Fire Victims Return to the Scene of the Disaster," *Los Alamos Monitor*, reprinted from May 16, 2000 in *Cerro Grande, Facing the Flames*, June 18, 2000, p. 24.
12. Shonda Novak, "Blaze Ignites Outrage in Los Alamos," *The New Mexican*, Sante Fe, New Mexico, May 10, 2000.
13. Roy Weaver, "Full text of Weaver's Apology," *Albuquerque Journal*, July 13, 2000. Online: www.abqjournal.com/news/1apology07–13–00.htm
14. Stephen J. Pyne, "No Fuel Like an Old Fuel; or, The Perils of Prescribed Burning," May 18, 2000. Online: http://www.public.asu.edu/spyne/nofuel.htm
15. U.S. National Park Service, "Cerro Grande Prescribed Fire Investigation Report," executive summary, May 18, 2000. http://www.nps.gov/cerrogrande/
16. Robert G. Stanton, Director, National Park Service, "Statement before the Senate Committee on Energy and Natural Resources," July 27, 2000. Online: http://www.doi.gov/ocl/2000/cerro.htm
17. Barry T. Hill, *Fire Management, Lessons Learned from the Cerro Grande (Los Alamos) Fire*, GAO/T-RCED-00–257. (Washington, D.C.: General Accounting Office, 2000).
18. Allen, "Panel Discussion."
19. Gleason, "Panel Discussion."
20. Hill, *Fire Management*, 11.
21. John Robertson, Fire Behavior Analyst, "Cerro Grande Prescribed Fire Investigation Report," National Parks Service, May 4–8, 2000, Appendix 8. Online at: http//www.nps.gov/cerrogrande.htm
22. Jack Cohen, "Why Los Alamos Burned," *Forest* (2000). Internet: http://www.forestmag.org/why-la-burned.html.
23. Quoted in Tony David, "The West's Hottest Question: How to Burn What's Bound to Burn," *High Country News*, 32, no. 11, June 4, 2000.
24. Mat Johansen, "Did Some Plead with Weaver?" Letter to Editor, *Los Alamos Monitor*, May 11, 2001.
25. "Cerro Grande Prescribed Fire Board of Inquiry Final Report," National Park Service, February 26, 2001, p. 45. For Los Alamos newspaper coverage, see: "Report Finds More Fault with Policies Than People," *Los Alamos Monitor*, June 13, 2001; and "Udall Criticizes Fire Report," *Los Alamos Monitor*, June 14, 2001.
26. Transcript of NPR program online at: http://www.cei.org/Transcripts.asp?ID=1163
27. National Public Radio, "Talk of the Nation," August 14, 2000. Hosted by Juan Williams. Program transcript online: http://www.cei.org/Transcripts.asp?ID=1163
28. Public Law No. 106–291.

29. *Federal Register*, Vol. 65, No. 218, pp. 67,480–67,511; earlier cohesive strategy summary document is: Barry T. Hill, "Western National Forests—A Cohesive Strategy Is Needed to Address Catastrophic Wildfire Threats," GAO/RCED-99–65 (Washington, D.C.: General Accounting Office, 1999).
30. Editor, "Closely Watched Fires," *Deming Headlight*, New Mexico, reprinted in *Sacramento Bae*, July 23, 2000, p. I4.
31. Editor, "Fire Against Fire," *Los Angeles Times*, August 5, 2000, p. B9.

13. Peaceful Coexistence

> The antagonist is no longer fire suppression. The debate is internal, over the right ends, means, and places to reinstate fire. The choice is not between fighting fires and lighting them but over the proper ways and times to do each.
>
> Pyne, 2000

The twentieth century war against fire on America's wildlands has not ended. Steward Edward White's analogy, over 80 years ago, of a bear caught by its tail that cannot safely be released, has proven to be accurate. The "bear" has steadily grown more dangerous and powerful since White wrote about it in 1920. Today's wildfires are far bigger, far more destructive, and often impossible to defeat without the help of a weather change. With increasing numbers of people living at the edge of wildlands, there seems little choice, each fire season, but to muster the armies of fire fighters and equipment, pay the incredible costs, and fight to protect lives and property. During the 2000 fire season, costs exceeded $1.36 billion as 8.4 million acres burned.

Faced with those numbers and the aftermath of the Cerro Grande fire that burned into Los Alamos, the National Fire Plan was reviewed and updated again. "The task before us—reintroducing fire—is both urgent and enormous," the interagency review panel said. "Conditions on millions of acres of wildland increase the probability of large, intense fires beyond any scale yet witnessed. These severe fires will in turn increase the risk to humans, to property, and to the land upon which our social and economic well being is so intimately intertwined."[1]

The review team found that the 1995 fire policy was generally sound and appropriate, but in just five years conditions had further deteriorated and fire hazards were worse than had been previously understood. Wildland—urban interface risks, in particular, were now seen to be even more complex and extensive. The guiding principles for the updated National Fire Plan continued to give top priority to fire fighter and public safety. The role of fire as an essential ecological process and natural change agent was reaffirmed. Fire management plans to guide activities, both of wildfire suppression and fuel hazard reduction, were essential to the entire policy and were to be supported with the best available science.

Such findings and policies were overdue validation for the pioneers in prescribed burning, those early "torchbearers" who persisted in the face of resistance. Their scientific findings and philosophical arguments had finally been accepted. The nation had moved far beyond the debates of the early century and the controversies of the 1970s, when total fire management had theoretically become national policy. Yet, as enlightened as the new policies were, policy direction had not been translated into reality.

On July 31, 2001, a congressional oversight hearing heard testimony on the progress of the progressive, multi-billion dollar effort. The committee was told by the General Accounting Office that only 42 percent of the fire management plans necessary to comply with the 1995 policy were in place, and those plans encompassed only 31 percent of all acres of burnable vegetation.[2] Without the plans, there is no alternative to all-out fire suppression efforts. In 2001 the Department of Interior had targeted 1.4 million acres for hazardous fuels treatment and the Forest Service had hoped to treat 1.8 million acres. But by July 2001 Interior had only treated 515,000 acres, and the Forest Service was just up to 859,000 acres. Eighty percent of the acreage had been treated with prescribed fire; the rest mechanically or with hand labor.[3]

Of the 415 million acres of federal land that are considered "fire-adapted," about 200 million acres evolved with frequent fire (with fire return intervals under 35 years). Because of the Hundred Years War against nature's fire, 70 million of those acres were now in critical need of fuel reduction to restore ecological integrity and avoid future holocaust fires. Another 141 million acres were at moderate risk levels, a danger to public safety and ecological health. Those acres con-

Acres Treated with Prescribed Fire by Federal Agencies						
Agency	1995	1996	1997	1998	1999	2000
Forest Service	570,300	617,163	1,097,658	1,489,293	1,379,960	728,237
Bureau of Indian Affairs	21,000	16,000	37,000	48,287	83,875	3,353
Bureau of Land Management	56,000	50,000	72,500	200,223	308,000	125,600
National Park Service	62,000	52,000	70,000	86,126	135,441	19,072
Fish and Wildlife Service	209,000	180,000	324,000	285,758	300,508	201,052
Rx Fire Total	918,000	915,163	1,601,158	1,889,564	2,240,105	1,077,314

Wildfire acres	1995	1996	1997	1998	1999	2000
National Totals	2,315,730	6,701,390	3,672,616	2,329,709	5,661,976	8,422,237

Source: National Interagency Fire Center. *Statistics*. Http://www.nifc.gov/stats/

Prescribed Fire and Wildfire Statistics, 1995–2000.

tinued to degrade and, without treatment, would also become critical.[4]

"Up to now, utilizing controlled fires has been more of a curiosity than a significant, regularly applied national tool," former Forest Service Chief Jack Ward Thomas said. "This challenge from congress and the administration is in reality almost a leap of faith, and certainly the best chance in well over a decade for agencies to demonstrate that objectives can be formulated, constituencies constructed, and resulting programs executed."[5] Despite the "paradigm shift" that Thomas applauded, he was also disappointed that "not much has really changed" when he observed the 2001 wildfire season. "They're out there fighting fires as hard as they can, activating military units. They have to learn *how* to do this. They just can't hold themselves back when a fire is going."[6]

The bulk of the new multi-billion dollar budget was used to hire fire fighters and buy fire-fighting equipment. As the annual fire season closed down, an expanded effort at off-season fuel reduction using prescribed burns would take advantage of the additional staff and equipment. Fire fighters can, increasingly, become year-around fire technicians, able to conduct burns. But one of the obstacles that remained was the institutional approach to fire funding. Fuel reduction efforts are funded by fixed budgets, while fire suppression receives virtually unlimited emergency funding. Fire-fighting institutions and the large industry that markets fire equipment and supplies need to be weaned from profligate sums—a task that could prove as difficult as cutting the Pentagon's military-industrial budgets whenever a war has ended.

"On the ground, policy must contend with reality," seasonal wilderness ranger Emma Brown wrote. Her article in the *Washington Post* on April 29, 2001, was titled, "What Burns Me About the Way We Fight Wildfires." "The reality I experienced was that old attitudes persist, and old habits die hard. Longtime firefighters who love what they do are fierce about putting out flames, and they are used to having ample resources to do the job."

Part of the new budget was spent for prescribed burns on the Inyo National Forest in California in 2001. There, in the first week of May, a helicopter helped ignite a burn that cleaned logging slash and fine fuels on 900 acres of Jeffrey pine forest. About sixty firefighters worked this burn. Many of them had recently been hired for the coming fire season and were smelling their first smoke and feeling the heat thrown by well-behaved flames as the fire progressed. A veteran crew leader answered their questions about which trees might survive the scorching flames that curled around the base of the Jeffreys (close relatives to ponderosa pines). The helicopter, with a crew that goes to Florida each winter to burn in southern pine forests, carried a machine that injected "ping-pong balls" just before they were released in the air. The chemicals combined in a heat-releasing reaction that ignited into tiny bits of flame less than two minutes after the balls hit the ground. With the helicopter, the project coordinator expected costs for this burn to be under $25 per acre. Many prescribed burns cost around $100 per acre, but there are economies of scale when hundreds of acres are burned in a day. The bill for fighting most wildfires runs between $1,000 to $2,000 per acre.

The coordinator was gratified that the wind, as forecast, kept pushing smoke out over uninhabited wildlands and away from the nearest communities. Choosing the conditions for smoke management is one of the advantages prescribed burns have over uncontrolled fires. This project coordinator was conducting his first prescribed burn (with close supervision by experienced burn bosses). He had never heard of Harold Biswell or Harold Weaver.

Yet the principles governing the reintroduction of fire, using modern tools or not, were those perfected by the Harolds and later torchbearers who endured so much controversy. The obstacles that continue to hinder fire management suggest that some of the most important lessons today's burn specialists can learn from those men have to do with attitude and perseverance.

MUST WE RESTORE FIRE?

Today many questions about wildland fire begin with broad professional and public recognition that there is far too much hazardous fuel. Even that condition has been questioned by a few people who consider dense undergrowth as simply "natural" succession. Of course, that ignores the century of fire suppression that transformed wildlands. The forests that evolved with frequent fire do not look like and do not function as they once did. The succession of shade tolerant firs beneath ponderosa pine is the result of human intervention that now endangers once-stable old-growth forests (even without any threat of logging).

Today's hazard reduction efforts recognize that lives and property are at risk. Beyond self-preservation, the additional scientific and philosophical understanding that many landscapes are now "starved" of fire has not spread widely enough through American society. Simply reducing the mass of dead fuel mechanically may be possible (if done properly), and some of the fertilization when fire recycles nutrients may be chemically duplicated, if we choose to go that route. But there are heat effects that mechanical manipulation and chemical fertilizers cannot replicate. And fire moves across a landscape with random variability that transforms all of its impacts into ecosystem biodiversity for soil, plants, and animals, that makes every forest far "more than the sum of its cellulose." Interior Secretary Bruce Babbitt has made that point, adding, "We must not sacrifice the integrity of creation at the altar of commercial timber production."[7]

"Emphasizing *just* fuel reduction is a recipe for disaster," University of Washington professor James Agee warns.[8] Jim Agee and Jan van Wagtendonk, who both received early training from Harold Biswell, sometimes sound frustrated by the slow pace of change. Issues that put brakes on the pace of prescribed burning have been called, by Agee, "fine filters" that can clog the "coarse filter" of fire. As Gifford Pinchot and other early professional foresters recognized, fire often is the major selection factor—or filter—that determines the components of an ecosystem.

One "fine filter" that has slowed the reintroduction of fire is the air quality issue. Many regions suffer from atmospheres choked by "people fumes"—the toxins of an industrial and automobile age that overloads air basins and makes breathing itself a health risk. Smoke from deliberately set fires must "compete" with the levels of ever-present pollutants authorized under air quality regulations. "The window of burning has become much narrower," Yosemite scientist Jan van Wagtendonk says. "[In the early years] we burned with impunity, practically, from the smoke regulations. What we produce with prescribed fires is nothing compared to what the automobile is bringing up on us, so it's a societal issue. A park or a forest is easy for an air pollution control district to point to, it's harder to point to 10,000 drivers of automobiles, so we're an easy target."[9]

Another example of "fine filters," pointed to by Agee, is the policy in northwest forests to limit ground disturbance to protect sensitive species. Both logging and fire have been "lumped together" as ground-disturbing activities. Although a choice to do no logging stops that activity, the choice to do no prescribed burning is, instead, "an active choice *to rely on* wildfire." Those wildfires, when they arrive, are typically more intense and cause more widespread impacts under today's fuel loads. "Fine filter strategies will backfire in the long run," Agee says. "Recovery after wildfires can take one hundred plus years—they are a hundred-year curse."

That curse may partly explain why 135 out of 146 threatened, endangered, and rare plant species in the lower 48 states benefit from wildland fire or are found in fire adapted ecosystems.[10]

"We thought it was as simple as this slight revision of Shakespeare's question, 'To burn or not to burn, that is the question,' " Jim Agee added. "But the question wasn't really that simple and the answers aren't either. It's not always our choice whether fire will visit a landscape. But in itself, that realization should help guide our future fire strategies."[11]

Looking back at his graduate student years with Biswell, Agee sees them as "simpler times than we see now. I kind of liken it to a computer where you keep adding programs and you're trying to run more things simultaneously and it just gets slower and slower and slower. When we started out, our 'computers' were much simpler but they weren't weighted down by so much baggage."[12]

Jan van Wagtendonk recognizes the progress made since the burn program began at Yosemite National Park in the 1970s, however, "the disappointment has come more recently, where the program has not continued at that same pace. Superintendents that don't want to burn; smoke regulations that don't allow you to burn; the whole timing thing after Yellowstone has not allowed us to keep pace. So it gets worse every year. We're not keeping up with it. There is hope, with the new emphasis now, that we can divert enough of that into prescribed burning to ramp up, but still we end up with the smoke restrictions."[13] Yet, the Yosemite researcher concluded, "I'm an optimist by nature. That is the sort of response that Doc [Biswell] gave me, too. When I asked him [a few years before his death in 1992], whether he was frustrated by the slow pace, he said, 'No! Look how far we've come!' I agree with Doc about that, but then, look how far we have yet to go."

Van Wagtendonk is one of many people warily watching where the emphasis goes and how the money is spent. "There's still a lot of hesitancy for the manager to use fire. In the federal program to increase fuels management, it did not specify that you're going to use prescribed fire to treat fuels; it said 'you're going to treat fuels.' So they're looking at mechanical and other treatments, because it's less risk for the manager to go out and cut this stuff than to light a match."[14]

LOGGING VERSUS THINNING

Along with the new comprehensive strategy for forest management and the funding for fuels reduction came debates about how best to accomplish those goals. Some of the arguments and questions seemed to ignore a century of relevant history. Timber companies saw the new initiatives as an opportunity to increase the logging "cut" on national forest lands. If too much wood was in the forests, it seemed intuitive, to some people, that cutting down trees must help the situation. Many pointed to the massive fires in the 1990s as evidence that not enough logging was going on.

Yet, throughout the century large fires had followed logging. Traditional timber harvesting took the least flammable material, the big trunks, off-site to be milled. The tops and branches and smaller, kindling-sized material remained behind. Slash disposal was one of the perennial challenges for the forestry profession, particularly in dry environments where little decomposition occurred. The incidence of human-caused fires also tended to follow logging roads, which served as access for recreation vehicles.

When a federal policy to stop building new roads within national forests was proposed, a logging industry spokesman told the media, "We are relegating those lands to blackened forest. Just as sure as we're talking here, those areas are going to burn up and we can't, without active management, reduce the seriousness of those fires."[15]

Countering such arguments were the findings of the Sierra Nevada Ecosystem Project, whose scientists reported to Congress that "timber harvest, through its effects on forest structure, local microclimate, and fuels accumulation, has increased fire severity more than any other recent human activity."[16] Pacific Northwest Research Station studies concluded, "Logged areas generally showed a strong association with increased rate of spread and flame length . . . positively correlated with the proportion of area logged. . . . Even though these hazards diminish, their influence on fire behavior can linger for up to 30 years."[17] Also, "the high rate of human-caused fires has generally been associated with high recreational use in areas of higher road densities."[18] Fires are almost twice as likely to occur in roaded areas as they are in roadless areas.

"Fires Not Result of Reduced Logging, Neutral Study Says," a *Chicago Tribune* headline read, on September 10, 2000. A bipartisan research group, the Congressional Research Service, concluded, "If anything, heavy logging from earlier years may have contributed more to the conditions that have made western forests ripe for big fires, because more flammable small trees and heavy brush are often left in the forest after the larger stands of timber have been taken out."

Agriculture Secretary Dan Glickman told newsmen that the majority of the wildfires in the year 2000 were in areas that had been logged in the past, "not in the uncut, roadless forests that the timber interests covet most, and therefore portray as imperiled." A *Casper Star Tribune* editorial, on September 1, added: "This week in Montana, three-fourths of the largest fires were in logged and developed areas—including all five of the blazes rated most destructive in terms of lost homes and structures." The editorial headline was: "Fire in the West—Forests Can't Be Logged Back to Health."

Rather than logging, what was needed was thinning—removing undergrowth that was not big enough to be commercially viable under traditional timber practices. Part of the 2001 federal funding was to foster markets for new commercial uses of smaller fuels. Increased use of chipped material to generate electricity was an option being considered.

The Grand Canyon Forests Partnership was pointed to as one potential model for other areas. The U.S. Forest Service and a local foundation aimed to restore natural ecosystem functions and at the same time reduce the risk of catastrophic fire. The program would thin 100,000 acres of ponderosa pine forests within Flagstaff, Arizona's, urban-wildland interface and then restore a low-intensity fire regime.[19] Critics of the Flagstaff program worried that they would overharvest big trees and that this one model for management might become an inappropriate "cookbook" recipe for other areas instead of being tailored to local conditions and needs.

Historian Stephen Pyne would like to see the debates concluded and more action taken toward the ultimate goal to return fire to ecosystems:

> If excess fuel is the problem, then remove it. Haul it off, burn it in fireplaces and powerplants, mulch it into compost, send it through woodchippers. Crush it, crop it. Browse it with goats. Thin and stack it before burning. Burn sun-dried cuttings while the surrounding woods are still green. Burn piles in the snow. A prescribed burn is not a vaccination, an on–off inoculation against conflagration. It involves instead a series of burns, often with complex preparations, then repeated to perpetuity. We may need to become fire foragers, with crews in a constant search for small niches of fuels, moving and burning; that requires lots of time in the field.[20]

Pyne's "fire forager" suggestion sounded much like the never-adopted 1968 plan prepared by Robert Barbee for a permanent crew of prescribed burn specialists working year-around in Yosemite National Park. That was an approach that Harold Biswell had advised and similar to the way state park ranger Glenn Walfoort, Biswell's acolyte at Calaveras Big Trees State Park, had worked. "Small crews can get a lot done," Walfoort told me. "As the program grew, they threw money at it; I didn't need all that money, all that management. It just made for more logistical problems." Calaveras is a 6,500-acre park. The effectiveness of crews may be improved when their responsibilities are focused on a smaller landscape scale than the millions of acres found in some national parks and forests. Parceling out responsibility for subunits, in those instances, may be one way to build in flexibility that is lost when big burns with big logistical problems are attempted.

One of the most successful private organizations using prescribed fire has been the Nature Conservancy. In the 1990s it averaged over

400 annual burns on 60,000 to 70,000 acres across the nation.[21] While each of its reserves had a small staff, its trained burn specialists traveled to each other's sites to carry out burns, much like a statewide burn team that California's park system ultimately developed.

Pyne's mention of burning patches in the snow also sounded like practices known among the Miwok Indians of Yosemite Valley. They ignited vegetation as snow melted and vegetation was exposed in the early spring. That is an echo of one question that was debated at the beginning of the century, the practices of "Paiute" or "Indian forestry."

HOW MUCH *DID* INDIAN FIRES INFLUENCE THE LANDSCAPE?

One philosophical question that has persisted through the century comes full circle in debates over "natural" versus "human ignitions" in wilderness, and over mechanical thinning versus prescribed burning to reduce fuel loads. Were America's wildlands, before Europeans arrived, "pristine wildernesses" untouched by man, or "gardens" maintained by Indians whose nurturing effect was lost?

In a book published at the end of the century, *America's Ancient Forests*, Dr. Thomas Bonnicksen asserted that "by the time Europeans set foot in North America, Paleoindians and their descendants may have as much as doubled the number of fires that would normally burn because of lightning."[22]

An analysis of that book for *Forest Magazine* quoted some environmentalists who were suspicious about a pro-logging agenda in suggestions that humans managed the ancient forests. Their questions sounded like an echo of "professional foresters' " statements one hundred years earlier. "The amount of burning that indigenous people could have done was minuscule," Scott Silver, executive director of Wild Wilderness, is quoted as saying. "There's no way that these primitive people could have done landscape management over major portions of the landscape."[23]

Can that question, at this late date, ever be resolved? Dr. Kat Anderson, in 1996, wrote that for most of the 12 million Native Americans in precontact North America, "primitive" hunting and gathering were just parts of a "comprehensive land-management system. Simply put, acorns get wormy, old berry bushes produce less fruit, fire-dependent mushrooms don't grow every year, bunchgrasses decline in productivity as they accumulate dead material, and meadows shrink as trees encroach on them—all processes that can be reversed by active management . . . to support human populations. The principal management tool was fire—millennia of prescribed burns.[24]

Bonnicksen supplied several first-person accounts, including this from a Cree-Metis elder from the Grande Cache area of the Rocky Mountains, that illustrate the Indians' knowledge of local wind and fuel conditions. They would avoid burning during the day when winds blew uphill, "a dangerous time for fires," and wait for cool mountain breezes that moved downhill: "We'd always wait until late afternoon and the fire was set at the upper end [of the meadow]. It would burn down to the low, damp places where the really wet grasses grow. That's the way we burned mountain meadows. See, you have to know the wind; you have to know how to use it."[25]

FACING THE RISKS

Knowledge reduces risk, yet risk remains one of the major obstacles slowing the implementation of the National Fire Plan. After forty years with the U.S. Forest Service, Bob Mutch retired in 1994. Now a private fire management consultant (his work, in 2001, took him to Italy, Ethiopia, and Mongolia), Mutch is particularly concerned about a double standard that impairs

> our ability to prescribe fire on the landscape on a large enough scale to sustain healthy systems. A prescribed fire can be well-planned and well-executed by qualified people, but the moment something starts to go awry the support from politicians and the public is quickly lost.
>
> In contrast, practically any professional strategy can be adopted in suppressing a wildfire and vast amounts of money can be spent in implementing that strategy. No matter how adverse the outcome . . . politicians and the public generally side with the firefighter. For example, in a Malibu neighborhood in 1993 where practically every house burned to the ground, the signs on the street said, "Thank you, firefighters." This double standard is part of our tradition and culture.

Mutch sees compelling beauty in the ability of one prescribed fire to self-regulate the extent and behavior of future fires. "If the poetry of prescribed fire is clear enough, politicians and the public may begin to accept the risks associated with prescribed fires, just as they do with the risks of wildfires."

Bruce Kilgore, who has been called "the father of natural fire in the National Park Service,"[26] retired in 1997. He also remains focused on the issue of risk. "My recent conclusions," he wrote, in December 2000,

> are that in all our efforts to be cautious and reasonable after the Yellowstone and Bandelier fires, we need to be careful not to suppress all ecologically significant fires in parks and wilderness. As fire managers and resource managers, and as members of the American public, we need to look again at the severe restrictions sometimes placed on prescribed natural fires and prescribed burns in wildlands.
>
> Managers must take reasonable calculated risks. We must find ways to reward, not just penalize, risk taking by state and federal land managers. In effect, we must encourage the maximum possible role for natural fire in park and forest wildlands—including high intensity fires—while still giving reasonable consideration to safety of human life and property.[27]

Harold Biswell had blamed overcautious concerns about risk on "misunderstanding and confusion about fire itself and its proper use." In his 1980 speech to the American Association for the Advancement of Science, he said, "This is due in part to such things as different intensities of fire, different vegetation types, different soils, not identifying whether a fire is wild or a prescribed fire, different objectives [and] expectations from fire use." The twenty-year-old predictions he made about the future of fire ecology proved to be accurately farsighted, considering recent incidents and the paradigm shift that has occurred in fire management.

> Wildfire damage and number of homes burned will become much greater. Suppression costs will continue to skyrocket. Prescribed burning will become better understood and more widely used. Silviculturists will learn that prescribed fires can be their most important tool in timber management and multiple-use forestry.
>
> Very slowly, but eventually emphasis on prescribed burning will replace that on fire prevention and suppression. When it does, fire suppression will become more effective and much cheaper. There will not be as much devastation from wildfires. However, that seems many years off.

> There will be a greater awakening to the fact that fire is natural and that we must use it for our own survival.
>
> We cannot expect to undo the failures of the past . . . in a short time. That is, neglecting to understand nature and working in harmony with it. We must exercise great patience, and persistence, too.[28]

There are other things that we all can do to help. As many Americans move out to the edges of wildlands, they *can* become more like the original Americans who paid close attention to the rhythms and processes of their neighboring landscapes. We all would benefit from a culture that values and fosters long-term thinking, that looks beyond today's choked forest to the healthy ecosystem that should be restored. We can recognize that we do not see forests, or forest fires, like those that the Native Americans knew. Much of our wild country and most of the fires that burn there have been transformed, by our actions, into something unnatural. Wherever the historic conditions have been restored so fires again fulfill a positive function, we should look beyond the black and ash of a still-smoking fire and see the process of regeneration that begins on that day.

We need to embrace a different, older relationship that humanity once had with fire. To recognize that fires are as essential to most of our wildlands as predators are essential to prey. We should recognize that we have been inundated, all of our lives, with sensationalist news stories and inaccurate fire prevention propaganda that makes it hard for us to overcome a conditioned fear and hatred against wildfire.

All of us can celebrate and honor fire management professionals, not just when they return blackened and exhausted from a fight with wildfire, but when they come home after finishing a routine prescribed burn. On those days they are not warriors, but peacemakers. In the long term, such routine tasks will be far more successful at protecting our property and lives.

We can accept the manageable smoke from prescribed burns to avoid the uncontrollable smoke events that come with holocaust fires. And we should pressure our government agencies to complete mandated fire management plans, fund prescribed fire programs, and hire and train the necessary professionals. There should be many experts with intimate knowledge of those landscapes that will need fire treatments in perpetuity.

If we do these things, we might also rejoin the long list of America's "fire-adapted species." The start of a new century is a proper time to end our twentieth century war against wildfire and seek peaceful co-existence with nature's fire.

NOTES

HB—Harold Biswell papers, held at the Brancroft Library, University of California, Berkeley (BANC MSS 2002/67 c).

1. U.S. Department of the Interior, "Review and Update of the 1995 Federal Wildland Fire Management Policy," report to the Secretaries of the Interior, of Agriculture, of Energy, of Defense, and of Commerce; the Administrator, Environmental Protection Agency; the Director, Federal Emergency Management Agency; and the National Association of State Foresters, by an Interagency Federal Wildland Fire Policy Review Working Group, Boise, Idaho, National Interagency Fire Center, January 2001, ch. 1, p. 2.
2. Barry T. Hill, witness statement before the Subcommittee on Forest and Forest Health, Committee on Resources, U.S. House of Representatives. Washington, D.C., General Accounting Officer, July 31, 2001. Online: www.housegov/resources/107cong/forests/2001july31/hill.htm
3. Acreage statistics from testimony in the oversight hearing of the House Subcommittee on Forest and Forest Health, July 31, 2001, by Dale Bosworth, Chief, USDA, Forest Service, and Tim Hartzell, Office of Wildland Fire Coordination, U.S. Department of the Interior. Online: www.House.gov/resources/107cong/forests/2001july31/
4. Ibid., ch. 1, pp. 7, 8.
5. Jack Ward Thomas, keynote speech at National Fire Ecology Conference, San Diego, California, November 28, 2000.
6. Jack Ward Thomas, personal communication, September 1, 2001.
7. Bruce Babbitt, "Fight Fire with Fire," Address to Commonwealth Club, San Francisco, September 1, 1998. Online: http://www.wildfirenews.com/fire/articles/babbitt.html
8. James Agee, Keynote Talk at Fire Ecology Conference 2000: The First National Congress on Fire Ecology, Prevention and Management, San Diego, California, November 27, 2000. Proceedings not yet published; personal record.
9. Jan van Wagtendonk, personal communication, November 29, 2000.
10. "Review and Update," ch. 1, p. 6.
11. Agee, Keynote Talk at Fire Ecology Conference 2000.
12. Jim Agee, personal communication, February 1, 2001.
13. Jan van Wagtendonk, personal communication, November 29, 2000.
14. Ibid.
15. Phil Aune, California Forestry Association spokesman, in: Robert A. Rosenblatt and Rebecca Trounson, "New Rule to Ban Building of National Forest Roads Policy," *Los Angeles Times*, January 5, 2001.

16. Quoted in Matthew Koehler, "The Truth about Logging and Wildfire Prevention." Wild Rockies Organization, 2000, p. 4. Online: http://www.wildrockies.org/wildfire.

17. U.S. Department of Agriculture, "Historical and Current Forest Landscapes in Eastern Oregon and Washington, Part II: Linking Vegetation Characteristics to Potential Fire Behavior and Related Smoke Production," PNW–GTR–355, Pacific Northwest Research Station, Portland, Oregon, 1994.

18. U.S. Department of Agriculture, "An Assessment of Ecosystem Components in the Interior Columbia Basin and Portions of the Klamath and Great Basins" Vol. II, PNW-GTR-405, Pacific Northwest Research Station, Portland, Oregon, 1997.

19. Grand Canyon Forest Partnership webpage: www.gcfp.org

20. Stephen J. Pyne, "No Fuel Like an Old Fuel; or, The Perils of Prescribed Burning," May 18, 2000. Online: http://www.public.asu.edu/spyne/nofuel.htm

21. The Nature Conservancy prescribed fire webpage: http://www.tncfire.org/vol8no1.htm

22. Thomas Bonnicksen, *America's Ancient Forests: From the Ice Age to the Age of Discovery* (New York: John Wiley, 2000), 147.

23. Matt Rasmussen, "The Long Reach of Humanity," *Forest Magazine* (March/April. 2000): 18.

24. M. Kat Anderson, "Tending the Wilderness." *Restoration & Management Notes* 14, no. 2 (Winter 1996): 158.

25. Bonnicksen, *America's Ancient Forests*. 202.

26. Conrad Smith, *Media and Apocalypse: News Coverage of the Yellowstone Forest Fires*, Exxon Valdez *Oil Spill, and Loma Prieta Earthquake* (Westport, CT: Greenwood Press 1992), notes, p. 71.

27. Bruce Kilgore, personal communication, December 28, 2000. Kilgore retired on March 31, 1997. He was presented the Department of Interior's Meritorious Service Award in 1993 and its Distinguished Service Award in 1999, in part because of his "pioneering achievements in fire ecology and prescribed fire practices" and his work that "influenced other agencies to adopt these practices."

28. Harold Biswell, "Fire Ecology: Past, Present, and Future," keynote talk to the Ecology Section, American Association for the Advancement of Science, Davis, California, June 23, 1980, pp. 7–9, speech notes, HB.

Selected Bibliography

Adams, Charles. "Ecological Conditions in National Forests and in National Parks." *The Scientific Monthly* 20 (June 1925): 561–590.

Agee, James K. *Fire Ecology of Pacific Northwest Forests*. Washington, D.C.: Island Press, 1993.

———. "Perceptions and Professionals: Coming to Grips with Both." *Renewable Resources Journal*. 11, no. 1 (Spring 1993): 25–26. In special report "Workshop on National Parks Fire Policy: Goals, Perceptions, and Reality."

———. "Memorial Dedication to Dr. Harold H. Biswell." In *The Biswell Symposium: Fire Issues and Solutions in Urban Interface and Wildland Ecosystems, February 15–17, 1994*. USDA Forest Service Gen. Tech. Rep PSW-GTR-158. Walnut Creek, California, Pacific Southwest Field Station, 1995, pp. 1–3.

Allen, C. D., R. Touchan, and T. W. Swetnam. "Landscape-Scale Fire History Studies Support Fire Management Action at Bandelier." *Park Science* 15, no. 13 (1995): 18–19.

Anderson, M. Kat. "Tending the Wilderness." *Restoration & Management Notes* 14, no. 2 (Winter 1996): 154–166.

Anderson, M. Kat, and Michael J. Moratto. "Native American Land-Use Practices and Ecological Impacts." In *Sierra Nevada Ecosystem Project Final Report to Congress*, Vol. II, *Assessments and Scientific Basis for Management Options*. Davis: University of California, Centers for Water and Wildland Resources, 1996.

Arno, Stephen F. "Forest Fire History in the Northern Rockies." *Journal of Forestry* (August 1980): 460–465

———. "History of Fire Occurrence in Western North America." *Renewable*

Resources Journal 11, no. 1 (Spring 1993): 12–13. In special report: "National Parks Fire Policy: Goals, Perceptions, and Reality."

———. "The Seminal Importance of Fire in Ecosystem Management—Impetus for This Publication." In *The Use of Fire in Forest Restoration*, General Tech. Report INT-GTR-341. USDA, Forest Service, Intermountain Research Station. June 1996.

———. "Eighty-eight Years of Change in a Managed Ponderosa Pine Forest." General Technical Report RMRS-GTR-23. USDA, Forest Service. Rocky Mountain Research Station. March 1999.

Ayres, R. W. "History of the Stanislaus." February 1911. Online at http://www.r5.fs.fed.us/stanislaus/heritage/voices/voices04.htm

Babbitt, Bruce. "Fight Fire With Fire." Address to Commonwealth Club, San Francisco, California. September 1, 1998. Online: http://www.wildfirenews.com/fire/articles/babbitt.html

Babbitt, Bruce, and Dan Glickman. "Managing the Impact of Wildfires on Communities and the Environment: A Report to the President in Response to the Wildfires of 2000." September 8, 2000. Online: http://www.whitehouse.gov/CEQ/fireport.html

Bancroft, Larry, Thomas Nichols, David Parsons, David Graber, Boyd Evison and Jan van Wagtendonk. "Evolution of the Natural Fire Management Program at Sequoia and Kings Canyon National Parks." Paper presented at the Wilderness Fire Symposium, Missoula, Montana, November 15–18, 1983, pp. 174–180.

Barbee, Robert D. "Replies from the Fire Gods." *American Forests* (March/April 1990): 34–35, 70.

Barrett, Louis A. *Record of Forest and Field Fires in California*. USDA Forest Service, California Region. San Francisco, 1935.

Barry, W. James, and R. Wayne Harrison. "Prescribed Burning in the California State Park System." Presented at the Symposium on Fire in California Ecosystems: Integrating Ecology, Prevention, and Management, San Diego, California, November 17–20, 1997.

Baskin, Yvonne. "Yellowstone Fires: A Decade Later; Ecological Lessons Learned in the Wake of the Conflagration." *BioScience* (February 1999). Online: http://www.findarticles.com/cf_0/m1042/2-49/53889669.

Biswell, Harold. "Prescribed burning in Georgia and California Compared." *Journal of Range Management* 11, no. 6 (1958): 293–298.

———. "Man and Fire in Ponderosa Pine." *Sierra Club Bulletin* 44, no. 7 (1959):44–53.

———. "Reduction of Wildfire Hazard." *California Agriculture* 13, no. 6 (1959):5.

———. "Danger of Wildfires Reduced in Ponderosa Pine." *California Agriculture* 4, no. 10 (1960): 5–6.

———. "The Big Trees and Fire." *National Parks Magazine*, April 11–14, 1961.

———. "Research in Wildland Fire Ecology in California." In *Proceedings, 1st Tall Timbers Fire Ecology Conference, March 1–2, 1962, 63–97*. Tallahassee, FL: Tall Timbers Research Station, 1963.

———. "The Use of Fire in Wildland Management." In *Natural Resources, Quality and Quantity*, edited by S. V. Wantrup and James J. Parsons. Berkeley: University of California Press, 1967.

———. "Fire Ecology in Ponderosa Pine-grassland." In *Proceedings, Annual Tall Timbers Fire Ecology Conference, June 8–9, 1971*, 69–96. Tallahassee, FL: Tall Timbers Research Station, 1972.

———. "The Role of Fire in Maintaining Forest Wilderness Quality." Paper presented at the Second Annual California Plant and Soil Conference, California Chapter, American Society of Agronomy. February 1, 1973.

———. "Some Aspects of Simulated Natural Fires in Vegetation Management." Society for Range Management. Omaha, Nebraska February 19, 1976. Session introductory comments.

———. "Prescribed Fire as a Management Tool." Paper presented at the Symposium on Environmental Consequences of Fire and Fuel Management in Mediterranean Ecosystems. Palo Alto, California, August 1–5, 1977.

———. "Fire Ecology: Past, Present, and Future." Keynote talk to the Ecology Section, American Association for the Advancement of Science, Davis, California, June 23, 1980.

———. *Prescribed Burning in California Wildlands Vegetation Management*. 1989. Reprint. Berkeley: University of California Press, 1999.

Biswell, Harold, and J. E. Foster. "Forest Grazing and Beef Cattle Production in the Coastal Plain of North Carolina." North Carolina Agriculture Experiment Station Bulletin No. 334, 1942.

Biswell, Harold H., Harry R. Kallander, Roy Komarek, Richard J. Vogl, and Harold Weaver. "Ponderosa Fire Management. Miscellaneous Publication No. 2." Tallahassee, Florida: Tall Timbers Research Station, 1973.

Boerker, Richard H. "Light Burning Versus Forest Management in Northern California." *Journal of Forestry* 10 (1912): 184–194.

Bolgiano, Chris. "Yellowstone and the Let-Burn Policy." *American Forests* (January/February 1989): 22–25, 74–78.

Bonnicksen, Thomas M. "Fire Gods and Federal Policy." *American Forests* (July/August 1989): 14–16, 66–68.

———. *America's Ancient Forest: From the Ice Age to the Age of Discovery*. New York: John Wiley, 2000.

Boyd, Robert. *Indians, Fire and the Land in the Pacific Northwest*. Corvallis, OR: Oregon State University Press, 1999.

Brennan, Leonard A., and Sharon M. Hermann. "Prescribed Fire and Forest Pests: Solutions for Today and Tomorrow." *Journal of Forestry* (November 1994): 34–36.

Bruce, Donald. "Light Burning—Report of the California Forestry Committee." *Journal of Forestry* 21 (1928): 129–133.

Butler, Mary Ellen. *Prophet of the Parks, The Story of William Penn Mott, Jr.* Ashburn, VA: National Recreation and Park Association, 1999.

Cammerer, Arno B. "Outdoor Recreation—Gone with the Flames." *American Forest* April (1939): 182.

Caprio, Anthony C., and David M. Graber. "Returning Fire to the Mountains: Can We Successfully Restore the Ecological Role of Pre-Euroamerican Fire Regimes to the Sierra Nevada?" In David N. Cole and Stephen F. McCool, *Proceedings: Wilderness Science in a Time of Change*. Proc. RMRS-P-000. USDA, Forest Service, Rocky Mountain Research Station. Ogden, Utah, 2000.

Cermak, Robert W. "Fire Control in the National Forests of California, 1898–1920." Master's thesis, California State University, Chico, 1986.

"Cerro Grande . . . Facing the Flames." Cerro Grande Fire Special Edition. *Los Alamos Monitor*, June 18, 2000.

Chapman, H. H. "Editorials: Fire—Master or Servant." *Journal of Forestry* 37 (1931): 605.

———. "Some Further Relations of Fire to Longleaf Pine." *Journal of Forestry* 30 (1932): 602–604.

———. "Fire and Pines . . . A Realistic Appraisal of the Role of Fire in Reproducing and Growing Southern Pines." *American Forests* 50 (1944): 62–64.

———. "Prescribed Burning Versus Public Forest Fire Services." *Journal of Forestry* 45 (1947): 804–808.

———. "Prescribed Burning in the Loblolly Pine Type." Exchange of letters with William L. Hall. *Journal of Forestry* 45 (1947): 209–212.

———. "To Whom It May Concern." *American Forests* 62, no. 8 (August 1956): 54–55.

Chapman, H. H., and H. N. Wheeler. "Controlled Burning." *Journal of Forestry* 39 (1941): 886–891.

Christensen, N. L., L. Cotton, T. Harvey, R. Martin, J. McBride, P. Rundel, and R. Wakimoto. "Review of Fire Management Program for Sequoia-Mixed Conifer Forests of Yosemite, Sequoia and Kings Canyon National Parks." Final report to National Park Service. Washington, D.C., 1987.

Christensen, Norman L., James K. Agee, Peter F. Brussard, Jay Hughes, Dennis H. Knight, G. Wayne Minshall, James M. Peek, Stephen J. Pyne, Frederick J. Swanson, Jack Ward Thomas, Stephen Wells, Stephen E. Williams and Henry A. Wright. "Interpreting the Yellowstone Fires of 1988." *BioScience* 39, no. 10 (November 1989): 678–685.

Clar, C. Raymond. *California Government and Forestry: From Spanish Days until the Creation of the Department of Natural Resources in 1927*. Sacramento: California State Board of Forestry, 1959.

———. "The Development of a Forest Fire Protection System in the California Division of Forestry, 1930–42." An interview by Mrs. Amelia Fry. Regional Oral History Office, Berkeley, University of California, May 29, 1966.

Cohen, Jack D. "Why Los Alamos Burned." *Forest Magazine* (2000). http://forestmag.org/why-la-burned.html.

Coman, Warren E. "Did the Indian Protect the Forest?" *Pacific Monthly* 26, no. 3 (September 1911): 300–309.

Conarro, R. M. "Fighting Tomorrow's Fires Today." *American Forests* (April 1939): 214.

Covington, W. Wallace. "Ponderosa Ecosystem Restortion and Conservation." Ecological Restoration Institute, Northern Arizona University School of Forestry, 1999. Online: http://www.eri.nau.edu/cov99vfn.html

Cowles, Raymond B. "Starving the Condor." *California Fish and Game* 44 (1958): 175–181.

———. "Fire Suppression, Faunal Changes and Condor Diets." In *Proceedings, Tall Timbers Fire Ecology #7: 217–224*. Tallahassee, FL: Tall Timbers Research Station, 1967.

Crosby, Bill. "Our Wild Fire: History Shows that Nearly All of California Is Designed to Burn." *Sunset* (June 1992): 64–72.

Daniels, Orville L. "Test of a New Land Management Concept: Fritz Creek 1973." *Western Wildlands* 1, no. 3 (1974): 23–26.

———. "Fire Management Takes Commitment." In *Proceedings Tall Timbers Fire Ecology Conference and Fire and Land Management Symposium, October 8–10, 1974*. Tallahassee, FL: Tall Timbers Research Station, 1976.

———. "A Forest Supervisor's Perspective on the Prescribed Natural Fire Program." In *Proceedings, 17th Tall Timbers Fire Ecology Conference, May 18–21, 1989. High Intensity Fire in Wildlands*. Tallahassee, FL: Tall Timbers Research Station, 1989.

David, Tony. "The West's Hottest Question: How to Burn What's Bound to Burn." *High Country News* 32, no. 11, June 5, 2000.

Dawson, Kerry J., and Steven E. Grego. "The Visual Ecology of Prescribed Fire in Sequoia National Park." In *Proceedings of the Symposium on Giant Sequoias: Their Place in the Ecosystem and Society, June 23–25, 1992*. Visalia, California. USDA Forest Service Gen. Tech Rep. PSW-151: 99–107, 1994.

DeBuys, William. "Los Alamos Fire Offers a Lesson in Humility." *High Country News*. 32, no. 13, (July 3, 2000).

"A Defense of Forest Fires." *Literary Digest*. August 9, 1913.

Demmon, E. L. "Fires and Forest Growth." *American Forests* 35 (April 1929).

Despain, Don G., and William H. Romme. "Ecology and Management of High-Intensity Fires in Yellowstone National Park." In *Proceedings, 17th Tall Timbers Fire Ecology Conference, May 18–21, 1989. High Intensity Fire in Wildlands*. Tallahassee, FL: Tall Timbers Research Station, 43–57.

Devlin, Sherry. "Check with Reality: Intense Blazes of 2000 May Be Wake-up Call to Return Fire to Forests." *The Missoulian*. August 22, 2000, Missoula, Montana. Online at: http://www.fs.fed.us/rm/main/pa/newsclips/00_08/082200_reality.html.

Doxey, Wall. "Fire or Forestry—The South's Great Problem." By representative from Congress from Mississippi. *American Forests* (April 1939): 161.

duBois, Coert. *Systematic Fire Protection in the California Forests*. USDA, Forest Service. Washington, D.C.: Government Printing Office, 1914.

———. "Cooperative Brush-Burning in the California National Forests." Draft circular, 95–97–03, Box 23 (27837) "Fire, Coop. 1915–23." NARA, San Bruno, California, 1915.

Easthouse, Keith. "Los Alamos Inferno: Los Alamos National Laboratory Has Long Played with Nuclear Fire. But Can It Handle a Forest Fire?" *Forest Magazine* (September/October 1999). Online at: http://www.forestmag.org/losalamosfire.htm

Egleston, H. H. *Report on the Relation of Railroads to Forest Supplies and Forestry*. U.S. Department of Agriculture, Forestry Division, Bulletin No. 1. Washington, D.C.: U.S. Government Printing Office, 1887.

Elfring, Chris. "Yellowstone: Fire Storm over Fire Management." *BioScience*. 39, no. 10 (November 1989): 667–672.

Evans, C. F. "Can the South Conquer the Fire Scourge?" *American Forestry* 50 (May 1944): 227–229.

"The Father of Smokey Bear Speaks." *Forest Log* (September–October 1994): 14–17.

"The Fire Next Time." *Time*, August 7, 1972, pp. 48–49.

Folweiler, A. D. "The Place of Fire in Southern Silviculture." *Journal of Forestry* 50 (1952): 187–190.

Gabrielson, Ria N. "Burning Wildlife." *American Forests* (April 1939): 186.

Gannett, Henry, ed. "Report of the National Conservation Commission, February, 1909; Special Message from the President of the United States." Washington, D.C.: Government Print Office, 1909.

Gillette, Charles A. "Campaigning Against Forest Fires." *American Forests* (April 1931): 209, 256.

Graves, Henry S. *Protection of Forests from Fire*. U.S. Department of Agriculture, Forest Service, Bulletin No. 82. Washington, D.C.: Government Printing Office, 1910.

———. "National Forests and National Parks in Wildlife Conservation."

Proceedings of the National Parks Conference. Washington, D.C.: Government Printing Office, 1917. Online at: http://memory.loc.gov/cgi-bin/query/D?consrv:5:./temp/~ammem_ZEVP

———. "D-5, Fire Cooperation Brush Burning." Letter to Coert duBois. 95–97–03, Box 23, "Fire, Coop. 1915–23." NARA, San Bruno, California, January 24, 1918.

———. "The Torch in the Timber: It May Save the Lumberman's Property, But It Destroys the Forests of the Future." *Sunset, the Pacific Monthly*. 44 (April 1920): 37–40, 80–90.

Graves, Walter L., and Gary Reece. "The Legacy of Harold Biswell in Southern California: His Teaching Influence on the Use of Prescribed Fire." USDA Forest Service Gen. Tech. Rep. PSW-GTR-158. First presented at the Biswell Symposium: Fire Issues and Solution in Urban Interface and Wildland Ecosystems, February 15–17, 1994, Walnut Creek, California, 1995.

Greeley, William B. "Austin Cary as I Knew Him." *American Forests* (May 1955): 30.

Green, Lisle R. *Burning by Prescription in Chaparral*. PSW-51. USDA-Pacific Southwest Forest and Range Experiment Station, Berkeley, California, May 1981.

Greene, S. W. "The Forest that Fire Made." *American Forests* (October 1931): 583–584, 683.

Gruell, George E. "Indian Fires in the Interior West: A Widespread Influence," 68–74. Presented at the Wilderness Fire Symposium, Missoula, Montana, November 15–18, 1983.

———. "A Prerequisite for Better Public Understanding of Fire Management Challenges," 25–38. *In Proceedings, 17th Tall Timbers Conference, May 21, 1989. High Intensity Fire in Wildlands*. Tallahassee, FL: Tall Timbers Research Station, 1989.

Guth, A. Richard, and Stan B. Cohen. *Red Skies of '88*. Missoula, MT: Pictorial Histories Publishing Co., 1989.

Hampton, H. Duane. *How the U.S. Cavalry Saved Our National Parks*. Bloomington: Indiana University Press, 1971.

Hanson, Chad. "The Big Lie: Logging and Forest Fires." *Earth Island Journal* 15, no 1 (Spring 2000). Online: http://www.earthisland.org/eijournal/spr2000/eia_spr2000jmp.html

Harper, Roland M. "A Defense of Forest Fires." *Literary Digest* (August 1913): 9.

Heinrichs, Jay. "The Ursine Gladhander." *Journal of Forestry* (October 1982): 642.

Hester, Eugene. "The Evolution of Park Service Fire Policy." *Renewable Resources Journal* 11, no. 1 (Spring 1993): 14–15. In Special Report: "Workshop on National Parks Fire Policy: Goals, Perceptions, and Reality."

Heyward, Jr., Frank. "History of Forest Fires in The South." *Forest Farmer* 9 no. 8 (1950): 3, 10–11.

———. "Austin Cary, Yankee Peddler in Forestry." Part 2 (first part in May 1955 issue). *American Forests* 62 (June 1955): 28.

Hill, Barry T. *Western National Forests—A Cohesive Strategy Is Needed to Address Catastrophic Wildfire Threats*. GAO/RCED-99–65. Washington, D.C.: General Accounting Office, 1999.

———. *Fire Management, Lessons Learned from the Cerro Grande (Los Alamos) Fire*. GAO/T-RCED-00–257 Washington, D.C.: General Accounting Office, 2000).

Holbrook, Stewart H. *Burning an Empire: The Story of American Forest Fires* New York: Macmillan Company, 1943.

Hoxie, George L. "How Fire Helps Forestry: The Practical vs. the Federal Government's Theoretical Ideas." *Sunset* 34 (August 1910): 145–151.

Hurley, Jerry. "Prescribed Burning in the 21st Century." USDA Forest Service Gen. Tech. Rep. PSW-GTR-158. 1995. Abbreviated version presented at the "Biswell Symposium: Fire Issues and Solutions in Urban Interface and Wildland Ecosystems," Walnut Creek, California, February 15–17, 1994.

Jenks, Cameron. *The Development of Government Forest Control in the United States*. Baltimore, MD: Johns Hopkins Press, 1928.

Jepson, W. L. "The Fire-Type Forest of the Sierra Nevada." *The Intercollegiate Forestry Club Annual* 1, no. 1 (1921): 7–10.

Johnson, K. Norman, John Sessions, Jerry Franklin, and John Gabriel. "Integrating Wildfire into Strategic Planning in Sierra Nevada Forests." *Journal of Forestry* (January 1998): 42–49.

Johnson, Von J. "Prescribed Burning: Requiem or Renaissance?" *Journal of Forestry* 82, no. 2 (1984): 82–90.

Kay, Charles E. "Aboriginal Overkill and Native Burning: Implications for Modern Ecosystem Management." Online: http://wings.buffalo.edu/academic/department/anthropology/documents/burning

Keifer, MaryBeth. "Fuel Load and Tree Density Changes Following Prescribed Fire in the Giant Sequoia-mixed Conifer Forest: The First 14 Years of Fire Effects Monitoring," 306–309. *Tall Timbers Fire Ecology Conference 20th Proceedings, Fire in Ecosystem Management: Shifting the Paradigm from Suppression to Prescription*. Tallahassee, FL: Tall Timbers Research Station, 1998.

Keifer, MaryBeth, Nathan L. Stephenson, and Jeff Manley. "Prescribed fire as the Minimum Tool for Wilderness Forest and Fire Regime Restoration: A Case Study from the Sierra Nevada, CA." In David N. Cole, and Stephen F. McCool. *Proceedings: Wilderness Science in a Time of Chance*. Proc. RMRS-P-000. Ogden, UT: USDA, Forest Service, Rocky Mountain Research Station, 2000.

Kennedy, Roger. "Fires Illuminate Our Illusions in the Southwest." *High Country News* 32, no. 13, July 3, 2000.

Kilgore, Bruce. "Research Needed for an Action Program of Restoring Fire to Giant Sequoias." *Intermountain Fire Research Council Symposium on "The Role of Fire in the Intermountain West."* 1970, 172–180.

———. "Restoring Fire to High Elevation Forests in California." *Journal of Forestry* 70 (1972): 266–271.

———. "The Ecological Role of Fire in Sierran Conifer Forests; Its Application to National Park Management." *Quaternary Research* 3 (1973): 496–513.

———. "Introduction—Fire Management Section," 7–9. Tall Timbers Fire Ecology Conference. October 8–10, 1974. Missoula, Montana 1976.

———. "Fire Management in the National Parks: An Overview," 45–57. Tall Timbers Fire Ecology Conference. October 8–10, 1974. Missoula, Montana, 1976.

———. 1976c. "From Fire Control to Fire Management; An Ecological Basis for Policies." In *Transactions North American Wildlife and Natural Resources Conferences* 41 (1976): 477–473.

Kilgore, Bruce, and H. H. Biswell. "Seedling Germination Following Fire in a Giant Sequoia Forest." *California Agriculture* (February 1971): 8–10.

Kitts, Joseph A. "Preventing Forest Fires by Burning Litter." *The Timberman* (July 1919): 91.

Koehler, Matthew. "The Truth About Logging and Wildfire Prevention." Wild Rockies Organization. Online: http://www.wildrockies.org/wildfire/

Komarek, E. V. "The Use of Fire: An Historical Background," 7. In *Proceedings First Annual Tall Timbers Fire Ecology Conference*. Tallahassee, FL: Tall Timbers Research Institute, 1962.

———. "Comments on the History of Controlled Burning in the Southern United States." In *Proceedings, 17th Annual Arizona Watershed Symposium.* Arizona Water Commission Report No. 5. Phoenix, Arizona, September 19, 1973.

———. "A Quest for Ecological Understanding: The Secretary's Review, March 15, 1958–June 30, 1975." Miscellaneous Publications No. 5. Tallahassee, Florida, Tall Timbers Research Station, 1977.

———. "Reflections by E. V. Komarek, Sr.," 31. In *Proceedings, 17th Tall Timbers Fire Ecology Conference, May 18–21, 1989*. "High Intensity Fire in Wildland Management, Challenges and Options." Tallhassee, FL: Tall Timbers Research, Inc., 1991.

Kotok, E. I. "Fire, a Problem in American Forestry." *Scientific Monthly* 31 (1930): 450–452.

———. 1934. "Fire, a Major Ecological Factor in the Pine Region of California." Proc. Fifth Pacific Science Congress, Canada. Univ. of Toronto Press.

Landers, J. Larry. "About E. V. Komarek, Sr." In *17th Tall Timbers Fire Ecology Conference, May 18–21, 1989*, 3. *High Intensity Fire in Wildland Management*. Tallahassee, FL: Tall Timbers Research, Inc., 1991. pp. 3, 4.

Larson, G. B. "Whitaker's Forest." *American Forests* 72, no. 9 (1966): 22–25, 40–42.

Laut, Agnes C. "The Fire Protection of the U.S. Forest Service." *American Forestry* 19 (November 1913): 711.

Lawter, Jr., William Clifford. *Smokey Bear 20252, A Biography*. Alexandria, VA: Lindsay Smith Publishers, 1994.

Lee, Robert C. "Can Reason Suppress the Fire Demon?" In *Rangeland Fire Effects, A Symposium, November 27–29, 1984*, edited by Ken Sanders, Jack Durham, et al., 93–97. Boise: Bureau of Land Management/ University of Idaho, 1984.

Leiberg, John B. *Forest Conditions in the Northern Sierra Nevada, California*. Department of Interior, U.S. Geological Survey, Professional Paper No. 8. Washington, D.C.: Government Printing Office, 1902.

Leisz, Douglas R., and Carl C. Wilson. "To Burn or Not to Burn: Fire and Chaparral Management in Southern California." *Journal of Forestry*. (February 1980): 94–95.

Leopold, A. Starker, S. A. Cain, C. H. Cottam, Ira N. Gabrielson and T. L. Kimball. "The Leopold Committee Report: Wildlife Management in the National Parks." *American Forests* (April 1963): 32–35, 61–63.

"Let 'Em Burn." *Time*, October 28, 1974, p. 78.

Lewis, Henry T. *Patterns of Indian Burning in California: Ecology and Ethnohistory*. Ramona, CA: Ballena Press, 1973.

———. "Why Indians Burned: Specific Versus General Reasons," 75–80. Presented at the Wilderness Fire Symposium, Missoula, Montana, November 15–18, 1983.

"Light Burning on Pine Forests." *American Forestry*, 19, no. 10 (1913): 692.

Little, Charles. "Smokey's Revenge." *American Forests* 99 (May/June 1993): 24–25, 58–60.

———. *The Dying of the Trees*. New York: Penguin Books, 1995.

Lockmann, Ronald F. *Guarding the Forests of Southern California*. Glendale, CA: Arthur H. Clark Co., 1981.

Loeffelbein, Bob. "Smokey Hits 40!" *American Forests* (May 1984): 33.

Los Alamos National Laboratory. *A Special Edition of the SWEIS Yearbook Wildfire 2000*. LA-UR-00-3471. New Mexico: Los Alamos National Laboratory, August 2000.

Lyman, Chalmer K. "Our Choice—A Mild Singe or a Good Scorching." *Northwest Science*. 21, no. 3: (1947): 129–133.

MacCleary, Doug. "Understanding the Role the Human Dimension Has Played in Shaping America's Forest and Grassland Landscapes." *Eco-Watch*. February 10, 1994.

Maclean, John N. *Fire on the Mountain: The True Story of the South Canyon Fire*. New York: William Morrow, 1999.

Maclean, Norman. *Young Men & Fire*. Chicago: University of Chicago Press, 1992.

Manning, Richard. "Friendly Fire." *Sierra* (January/February 2001): 40–41, 110.

Manson, Marsden. "The Effect of the Partial Suppression of Annual Forest Fires in the Sierra Nevada Mountains," *Sierra Club Bulletin* 34 (January 1906): 22–24.

Maunder, Elwood R. (interviewer). "Voice from the South: Recollections of Four Foresters." Oral history interviews with Inman F. Eldredge, Walter J. Damtoft, Elwood L. Demmon, and Clinton H. Coulter. Forest History Society, Santa Cruz, California, 1977.

McBride, Joe R. "Managing National Parks." *Renewable Resources Journal* 11, no. 1 (Spring 1993): 24–25. In special report: "Workshop on National Parks Fire Policy: Goals, Perceptions, and Reality."

McCormick, W. C. "The Three Million." *American Forests* (August 1931): 479–480.

McDowell, John. "The Year They Firebombed the West." *American Forests* (May/June 1993): 22–23, 55.

McLaughlin, John S. "Restoring Fire to the Environment in Sequoia and Kings Canyon National Parks," 391–394. *Tall Timbers Fire Ecology Conference (12)*. Tallahassee, Florida, June 8–9, 1972.

McManus, Reed. "Twice Burned? The Los Alamos Fire Rekindles Debate Over Logging." *Sierra* (September/October 2000): 16, 17.

Minnich, Richard A. "Fire Mosaics in Southern California and Northern Baja California." *Science* 219 (March 18, 1983): 1287–1294.

———. *The Biogeography of Fire in the San Bernardino Mountain of California*. Berkeley: University of California Press, 1988.

———. "Landscapes, Land-use and Fire Policy: Where Do Large Fires Come From?" In J. M. Moreno, ed., *Large Forest Fires*. Leiden, the Netherlands: Backhuys Publishers, 1998, 133–158.

Minnich, Richard A., and Yue Hong Chou. "Wildland Fire Patch Dynamics in the Chaparral of Southern California and Northern Baja California." *International Journal of Wildland Fire* 7, no. 3 (1997): 221–248.

Moore, William R. "From Fire Control to Fire Management." *Western Wildlands* 1, no. 3 (1974): 11–15.

Morrison, Ellen Earnhardt. *Guardian of the Forest: A History of the Smokey Bear Program*. New York: Vantage Press, 1976.

———. *The Smokey Bear Story*. Alexandria, VA: Morielle Press, 1995.

Morrison, Micah. *Fire in Paradise: The Yellowstone Fires and Politics of Environmentalism*. New York: HarperCollins, 1993.

Muir, John. *The Mountains of California*. 1894. Reprint. New York: American Museum of Natural History and Doubleday, 1961.

Murkowski, Frank H. "Lessons Must Be Learned from Los Alamos Fire." Press Release from Senator Frank H. Murkowski, Alaska, chairman of Senate Committee on Energy and Natural Resources. July 27, 2000. Online: http://energy.senate.gov/press/releases/losalamos.firehearing.htm

Murphy, E.W. "California Pays the Red Piper." *American Forests* (April 1939): 202. Special edition titled: "Yours in Trust-We Must Protect It from Fire."

Murphy, James L., and Frank T. Cole. "Villains to Heroes: Overcoming the Prescribed Burner Versus Forest Firefighter Paradox," 17–22. In *Proceedings 20th Tall Timbers Fire Ecology Conference*. Tallahassee, FL: Tall Timbers Research, Inc., 1998.

Mutch, Robert W. "I Thought Forest Fires Were Black!" *Western Wildlands* 1, no. 3 (1974): 16–21.

———. "Understanding Fire as Process and Tool." Adapted from "Fire Management Today: Tradition and Change in the Forest Service." Presented at Society of American Foresters National Convention. Washington, D.C., September 28 to October 2, 1975.

———. "Fighting Fire with Prescribed Fire: A Return to Ecosystem Health." *Journal of Forestry* (November 1994): 31–33.

———. "Use of Fire as a Management Tool on the National Forests." Statement before the Committee on Resources, U.S. House of Representatives, Oversight Hearing, Sept. 30, 1997. Online: http://resourcescommittee.house.gov/105cong/fullcomm/sep30.97/nutch.htm

———. "Will We Be Better Prepared for the Fires of 2006?" *Bugle: Journal of the Rocky Mountain Elk Foundation* (March/April, 2001).

National Interagency Fire Center. "Prescribed Fire Statistics." National Interagency Fire Center. Boise, Idaho, 2000. Online at: http://www.nifc.gov/stats/prescribedfirestats.html

———. "Wildland Fire . . . 2000 . . . What's the Story?" National Interagency Fire Center. Boise, Idaho, 2000.

National Interagency Prescribed Fire Training Center website: http://fire.r9.fws.gov/pftc

National Research Institute. "How Can We Live with Wildland Fire?" California Communities Program, University of California, Davis, 1995.

Nelson, Robert H. *A Burning Issue: A Case for Abolishing the U.S. Forest Service*. New York: Rowman & Littlefield, 2000.

Oberle, Mark. "Forest Fires: Suppression Policy Has Its Ecological Drawbacks." *Science* 165 (August 1969): 568–571.

Oettmeier, W. M. "The Place of Prescribed Burning." *Forest Farmer* (May 1956): 6, 7, 18, 19.

Olmsted, Frederick E. "Fire and the Forest—The Theory of 'Light Burning'." *Sierra Club Bulletin* 8 (January 1911): 42–47.

———. "Forest Devastation: A National Danger and a Plan to Meet It." *Journal of Forestry* 17 (1919): 911–935. Letter of Transmittal, November 1.

Ostrander, H. J. "How to Save the Forests by Use of Fire." *San Francisco Call.* Letter to Editor. September 23, 1902, p. 6.

Parsons, David J. "Objects or Ecosystems? Giant Sequoia Management in National Parks." *Symposium on Giant Sequoias: Their Place in the Ecosystem and Society, June 23–25, 1992. Visalia, California.* Three Rivers, CA: Sequoia & Kings Canyon National Parks, 1992.

Parsons, David J., and Stephen J. Botti. "Restoration of Fire in National Parks." In *The Use of Fire in Forest Restoration.* General Technical Report INT-GTR-341. June 1996, pp. 29–31. Online: http://www.fs.fed.us/rm/pubs/int_gtr341/gtr341_4.html

Parsons, David J., and Jan W. van Wagtendonk. "Fire Research and Management in the Sierra Nevada National Parks." In *Ecosystem Management in the National Parks* edited by W. L. Halvorson and G. E. Davis. Tucson: University of Arizona Press, 1996, 25–48.

Pinchot, Gifford. "The Relation of Forests and Forest Fires." *National Geographic* 10 (1899): 393–403.

———. *The Fight for Conservation.* New York: Doubleday, Page and Co., 1910.

———. *Breaking New Ground.* 1947. Reprint. Washington, D.C.: Island Press, 1998.

Plummer, Fred G. *Forest Fires: Their Causes, Extent and Effects, with a Summary of Recorded Destruction and Loss.* USDA Forest Service-Bulletin 117. Washington, D.C.: Government Printing Office, 1912.

Pyne, Stephen J. *Fire in America: A Cultural History of Wildland and Rural Fire.* 1982. Reprint. Seattle: University of Washington Press, 1997.

———. *World Fire: The Culture of Fire on Earth.* New York: Henry Holt and Co., 1995.

———. *Introduction to Wildland Fire.* New York: John Wiley & Sons, 1996.

———. *Year of the Fires.* New York: Viking Penguin, 2001.

Rasmussen, Matt. "The Long Reach of Humanity." *Forest Magazine* (March/April 2000): 14–19.

Redington, Paul G. "What Is the Truth? Conclusion of the Light-burning Controversy." *Sunset, the Pacific Monthly* 44 (June 1920): 56–58.

Rice, Carol. "A Balanced Approach: Dr. Biswell's Solution to Fire Issues in Urban Interface and Wildland Ecosystems." USDA Forest Service Gen. Tech. Rep. PSW-GTR-158. First presented at the "Biswell

Symposium: Fire Issues and Solutions in Urban Interface and Wildland Ecosystems," February 15–17, 1994. Walnut Creek, California, 1995.

Rodgers III, Andrew Denny. *Bernard Eduard Fernow, A Story of North American Forestry*. Princeton, NJ: Princeton University Press, 1951.

Romme, William H., and Don G. Despain. "Historical Perspective on the Yellowstone Fires of 1988." *BioScience*. Vol 39 November (10): 695–699.

Schiff, Ashley L. *Fire and Water, Scientific Heresy in the Forest Service*. Cambridge, MA: Harvard University Press, 1962.

Schimke, Harry E., and Lisle R. Green. "Prescribed Fire for Maintaining Fuel-Breaks in the Central Sierra Nevada." Berkeley, California, Pacific Southwest Forest and Range Experiment Station, 1970.

Schultz, A. M., and H. H. Biswell. "Reduction of Wildfire Hazard." *California Agriculture* 10, no. 11 (1956): 4–5.

Sellars, Richard West. *Preserving Nature in the National Parks*. New Haven, CT: Yale University Press, 1997.

Shea, John P. "Our Pappies Burned the Woods." *American Forests* (April 1940). 159–174.

Shipek, Florence. "Kumeyaay Plant Husbandry: Fire, Water, and Erosion Management Systems." In *Before the Wilderness: Environmental Management by Native Californians*, edited by Thomas C. Blackburn and Kat Anderson. Menlo Park, CA: Ballena Press, 1993.

Shoemaker, Len. *Saga of a Forest Ranger; A Biography of William R. Kreutzer, Forest Ranger No. 1*. Boulder: University of Colorado Press, 1958.

Show, Stuart Bevier. "Personal Reminiscences of a Forester, 1907–1931." Written at the request of R. E. McArdle, Chief, U.S. Forest Service. Berkeley, California, 1955.

———. "National Forests in California; An Interview Conducted by Amelia Roberts Fry." University of California, Regional Cultural History Project, Berkeley, 1965.

Show, S. B. and E. I. Kotok. "Forest Fires in California, 1911–1920: An Analytical Study." USDA, Dept. Circular 243. Washington, D.C., February 1923.

———. "The Role of Fire in the California Pine Forest." USDA, Department Bulletin No. 1294. Washington, D.C., December 1924.

———. "Fire and the Forest (California Pine Region)." USDA, Department Circular 358. Washington, D.C., August 1925.

Smith, Conrad. *Media and Apocalypse: News Coverage of the Yellowstone Forest Fires*, Exxon Valdez *Oil Spill, and Loma Prieta Earthquake*. Westport, CT: Greenwood Press, 1992.

Society of American Foresters "Fire as a Tool in Forest Protection and Management." *Proceedings, Society of American Foresters, Southern Cal-*

ifornia Section, Annual Meeting, December 1, 1962, Oakland, California, 1962.

Steen, Harold K. *The U.S. Forest Service: A History*. Seattle: University of Washington Press, 1976.

Sterling, E. A. "Attitude of Lumbermen Toward Forest Fires." In *Yearbook of the United States Department of Agriculture, 1904*. Washington, D.C.: Government Printing Office, 1905, pp. 133–140.

Stoddard, H. L. *The Bobwhite Quail, Its Habits, Preservation and Increase*. New York: Scribner, 1931.

———. "Use of Fire in Pine Forests and Game Lands of the Deep Southeast." In *Proceedings, First Tall Timbers Fire Ecology Conference*, 31–42. Tallahassee, FL: Tall Timbers Research Institute, 1962.

Suckling, Kieran. "Fire & Forest Ecosystem Health in the American Southwest." Southwest Forest Alliance/Southwest Center for Biological Diversity, 1996. Online: http://www.sw-center.org/swcbd/papers/fire-prm.html

Sullivan, Margaret. *Firestorm! The Story of the 1991 East Bay Fire in Berkeley*. Berkeley, CA: City of Berkeley, 1993.

Sweeney, J. R., and H. H. Biswell. "Quantitative Studies of the Removal of Litter and Duff by Fire under Controlled Conditions." *Ecology* 42, no. 3 (1961): 572–575.

Task Force on California's Wildland Fire Problem. "Recommendations to Solve California's Wildland Fire Problem." California Department of Conservation. June 1972.

Taylor, Ron. "Fire in the Redwoods." *Westways* (August 1968): 36–37.

Thurmond, Jack. "Through 1930 with the Dixie Crusaders." *American Forests* (March 1930): 151.

Timbrook, Jan, John R. Johnson, and David D. Earle. "Vegetation Burning by the Chumash." In *Before the Wilderness: Environment Management by Native Californians*, edited by Thomas C. Blackburn and Kat Anderson. Menlo Park, CA: Ballena Press, 1993.

U.S. Bureau of Land Management. "Using Fire to Manage Public Lands." BLM National Office of Fire and Aviation. Boise, Idaho, 1997. Pamphlet.

———. "Burning Issues; An Interactive Multimedia Program." Joint project of the Bureau of Land Management and Florida State University. Florida State University, Tallahasse, Florida, 2000.

U.S. Department of Agriculture, Forest Service. "The True Story of Smokey Bear." Western Publishing Company, 1960. Comic book.

———. "Rx Fire!" Southwestern Region, USDA Forest Service, 1994. Pamphlet.

———. "Living With Fire." USDA Forest Service, Southwestern Region, 2000.

U.S. Department of the Interior. "Review and Update of the 1995 Federal

Wildland Fire Management Policy." Report to the Secretaries of the Interior, of Agriculture, of Energy, of Defense, and of Commerce; the Administrator, Environmental Protection Agency; the Director, Federal Emergency Management Agency; and the National Association of State Foresters, by an Interagency Federal Wildland Fire Policy Review Working Group. Boise, Idaho, National Interagency Fire Center, January 2001.

U.S. General Accounting Office. "Western National Forests: A Cohesive Strategy Is Needed to Address Catastrophic Wildfire Threats." GAO-RCED-99–65. Washington, D.C. April 2, 1999.

U.S. National Park Service. "The Yellowstone Fires: A Primer on the 1988 Fire Season." Yellowstone National Park, California, October 1, 1988.

———. "Cerro Grande Prescribed Fire Investigation Report." May 18, 2000. Online: http//www.nps.gov/cerrogrande.htm

———. "Cerro Grande Prescribed Fire Board of Inquiry Final Report." National Park Service. February 26, 2001. Online: http://www.nps.gov/fire/fireinfo/cerrogrande/reports/Board_report-feb26final.pdf.

Vale, Thomas. "The Myth of the Humanized Landscape: An Example from Yosemite National Park." *Natural Areas Journal* 18, no. 3 (1999): 231–236.

van Wagtendonk, Jan W. "Wilderness Fire Management in Yosemite National Park." In *Earthcare: Global Protection of Natural Areas*, edited by E. A. Schofield. Proceedings of the Fourteenth Biennial Wilderness Conference. Boulder, CO: Westview Press, 1978.

———. "The Role of Fire in the Yosemite Wilderness." Presented at the National Wilderness Research Conference, Fort Collins, Colorado, July 23–26, 1985.

———. "Park Goals and Current Fire Policy." In special report: "Workshop on National Parks Fire Policy: Goals, Perceptions, and Reality." *Renewable Resources Journal.* 11, no. 1 (Spring 1993): 19.

———. "Dr. Biswell's Influence on the Development of Prescribed Burning in California," 11–15. In USDA Forest Service Gen. Tech. Rep. PSW-GTR-158, *The Biswell Symposium: Fire Issues and Solutions in Urban and Wildland Ecosystems, February, 15–17, 1994*. Walnut Creek, California, 1995.

Vitas, George. *Forest and Flame in the Bible*. A Program Aid of the Cooperative Forest Fire Prevention Campaign sponsored by the Advertising Council, State Foresters, and the U.S. Department of Agriculture, Forest Service—PA-93. Reprinted December 1961.

Vogl, Richard J. "Comments on Controlled Burning." Tall Timbers Fire Ecology Conference (9), April 10–11, 1969, Tall Timbers Research Station, Tallahassee, Florida, pp. 1–4.

———. "Smokey's Mid-Career Crisis." *Saturday Review of the Sciences* 1, no. 2 (March 1973): 23–29.

Wagle, R. F., and Thomas W. Eakle. "A Controlled Burn Reduces the Impact of a Subsequent Wildfire in a Ponderosa Pine Vegetation Type." *Forest Science* 25, no. 1 (1979): 123–128.

Wahlenberg, W. G., S. W. Greene, and H. R. Reed. "Effects of Fire and Cattle Grazing on Longleaf Pine Lands, as Studied at McNeill, Miss." Technical Bulletin No. 683. Washington, D.C., U.S. Department of Agriculture, June 1939.

Walker, T. B. "T. B. Walker Expresses His Views on Conservation," *San Francisco Chronicle*, January 5, 1913, p. 56.

Weaver, Harold. "Fire as an Ecological and Silvicultural Factor in the Ponderosa-Pine Region of the Pacific Slope." *Journal of Forestry*. 41 (January 1943): 7–15.

———. "Fire as an Ecological Factor in the Southwestern Ponderosa Pine Forests." *Journal of Forestry* 49 (February 1951): 93–98.

———. "Fire as an Enemy, Friend, and Tool in Forest Management." *Journal of Forestry* 53 (July 1955): 499–504.

———. "Effects of Prescribed Burning in Ponderosa Pine." *Journal of Forestry* 55 (February 1957): 823–826.

———. "Implications of the Klamath Fires of September 1959." *Journal of Forestry* 59 (August 1961): 569–572.

———. "Fire and Management Problems in Ponderosa Pine." *3rd Annual Tall Timbers Fire Ecology Conference, April 9, 10, 1964*, 61–79. Tallahassee, FL: Tall Timbers Research Station, 1964.

Weldon, Leslie A. C. "The Use of Fire in Forest Restoration: Dealing with Public Concerns in Restoring Fire to the Forest." General Technical Report INT-GTR-341. USDA Forest Service, Rocky Mountain Research Station. Online: http://www.fs.fed.us/rm/pubs/int_ugtr341/gtr341_6.html

Wheeler, H. N. "Controlled Burning in Southern Pine." *Journal of Forestry* 42 (June 1944): 449.

Whelan, Robert J. *The Ecology of Fire*. Cambridge, MA: Cambridge University Press, 1945.

White, John R. "Letters to the *Times*: Scare Heads Mislead." *Los Angeles Times*. August 29, 1928. Letter to editor from Superintendent, Sequoia National Park.

White, M. E. "Report on the McCloud River Cooperative Burning Area." 95–97–03, Box 23, "Fire, Coop. 1915–23." NARA, San Bruno, California, July 23, 1916.

White, Stewart E. "Woodsmen, Spare Those Trees! Our Forests Are Threatened; a Plea for Protection." *Sunset, the Pacific Monthly* (March 1920): 23–26, 108–117.

———. "Getting at the Truth. Is the Forest Service Really Trying to Lay

Bare the Facts of the Light-Burning Theory?" *Sunset, the Pacific Monthly* (May 1920): 62, 80–82.

Williams, Ted. "Incineration of Yellowstone." *Audubon* (January 1989): 38–85.

Wilson, Carl C., and James B. Davis. "Forest Fire Laboratory at Riverside and Fire Research in California: Past, Present, and Future." General Technical Report PSW-105. Pacific Southwest Forest and Range Experiment Station. Berkeley, California, May 1988.

Wooley, H. E. "What Has Been Accomplished in Fire Protection on the National Forests." *American Forestry* 19 (November 1913).

Index

About the Author

DAVID CARLE was a state park ranger in California for 27 years. Before retiring in 2000, he was at the Mono Lake Tufa State Reserve, where he participated in the prescribed burn program at Mono Lake. He also taught biology at Cerro Coso Community College. Now a freelance writer, he is the author of *Drowning the Dream: California's Water Choices at the Millennium* (Praeger, 2000) and *Mono Lake Viewpoint* (1992).

ALVERNO COLLEGE LIBRARY

634.96
C278
DEMCO